T0401056

ADVANCED POWER SYSTEMS USING BITUMINOUS COAL

ENERGY SCIENCE, ENGINEERING AND TECHNOLOGY

Additional books in this series can be found on Nova's website under the Series tab.

Additional E-books in this series can be found on Nova's website under the E-book tab.

ADVANCED POWER SYSTEMS USING BITUMINOUS COAL

DANIEL A. LAKATOS
EDITOR

Nova Science Publishers, Inc.
New York

For permission to use material from this book please contact us:
Telephone 631-231-7269; Fax 631-231-8175
Web Site: http://www.novapublishers.com

NOTICE TO THE READER

The Publisher has taken reasonable care in the preparation of this book, but makes no expressed or implied warranty of any kind and assumes no responsibility for any errors or omissions. No liability is assumed for incidental or consequential damages in connection with or arising out of information contained in this book. The Publisher shall not be liable for any special, consequential, or exemplary damages resulting, in whole or in part, from the readers' use of, or reliance upon, this material. Any parts of this book based on government reports are so indicated and copyright is claimed for those parts to the extent applicable to compilations of such works.

Independent verification should be sought for any data, advice or recommendations contained in this book. In addition, no responsibility is assumed by the publisher for any injury and/or damage to persons or property arising from any methods, products, instructions, ideas or otherwise contained in this publication.

This publication is designed to provide accurate and authoritative information with regard to the subject matter covered herein. It is sold with the clear understanding that the Publisher is not engaged in rendering legal or any other professional services. If legal or any other expert assistance is required, the services of a competent person should be sought. FROM A DECLARATION OF PARTICIPANTS JOINTLY ADOPTED BY A COMMITTEE OF THE AMERICAN BAR ASSOCIATION AND A COMMITTEE OF PUBLISHERS.

Additional color graphics may be available in the e-book version of this book.

Library of Congress Cataloging-in-Publication Data

Advanced power systems using bituminous coal / editor, Daniel A. Lakatos.
    p. cm.
 Includes bibliographical references and index.
 ISBN 978-1-61324-670-2 (hardcover)
 1. Gas power plants. 2. Fuel cell power plants. 3. Integrated gasification combined cycle power plants. 4. Coal gasification. 5. Bituminous coal. I. Lakatos, Daniel A.
 TK1061.A38 2011
 553.2'4--dc23
                          2011016628

Published by Nova Science Publishers, Inc. † New York

CONTENTS

PREFACE

Today's energy situation has created a dilemma for coal use in the United States. On one hand, the environmental challenges of using coal appear formidable, particularly with growing concern over the impact of carbon dioxide emissions on global climate change. On the other hand, the projected demand for electricity, coupled with high fuel costs, presents a near-term opportunity for the greater use of coal to ensure energy security for America. This solution to coal's "Catch-22" can be achieved through technological advancements that enable coal-based energy plants to produce much needed electricity and fuels for secure and stable economic growth. This book examines the development of the technological pathway of advanced power systems using bituminous coal.

Chapter 1- The United States Department of Energy's (DOE) Strategic Center for Coal funds research and development (R&D) whose objective is to improve the efficiency and reduce the cost of advanced power systems. In order to evaluate the benefits of on-going R&D, Noblis utilized their energy systems analysis capabilities and conceptual computer simulation models to quantify the impact of successful federally-funded R&D on future power systems configurations.

Chapter 2- The United States Department of Energy's (DOE) Strategic Center for Coal funds research and development (R&D) with the objective to improve the efficiency and reduce the cost of advanced power systems. In order to evaluate the benefits of on-going R&D, Noblis utilized their energy systems analysis capabilities and Aspen Plus computer simulation models to quantify the impact of successful federally-funded R&D on future power systems configurations.

In: Advanced Power Systems using Bituminous Coal
Editor: Daniel A. Lakatos

ISBN: 978-1-61324-670-2
© 2011 Nova Science Publishers, Inc.

Chapter 1

A PATHWAY STUDY FOCUSED ON NON-CARBON CAPTURE ADVANCED POWER SYSTEMS R&D USING BITUMINOUS COAL - VOLUME 1[*]

United States Department of Energy

NETL VIEWPOINT

Background

Today's energy situation has created a dilemma for coal use in the United States. On one hand, the environmental challenges of using coal appear formidable, particularly with growing concern over the impact of carbon dioxide (CO_2) emissions on global climate change. This threatens coal's long-term future. On the other hand, the projected demand for electricity coupled with high fuel costs (particularly high oil prices and volatile natural gas prices) presents a near-term opportunity for the greater use of coal to ensure energy security for America. The solution to coal's "Catch-22" can be achieved through technological advancements that enable coal-based energy plants to produce much needed electricity and fuels for secure and stable economic growth while protecting the planet by preventing air pollution and greenhouse gas emissions. It is the development of this technological pathway that is the focus of this report.

Objective

The mission of the Department of Energy's (DOE) Clean Coal Program is to ensure the availability of ultraclean, abundant, low-cost domestic energy to fuel economic prosperity and strengthen energy security while enhancing environmental quality. The technologies

---

[*] This is an edited, reformatted and augmented version of United States Department of Energy publication DOE/NETL-2008/1337, dated October 16, 2008.

presented in this document describe the multi-year strategy that will enable the Advanced Power Systems, (Gasification and Advanced Turbines Programs) Fuel Cells, and Sequestration Research and Development (R&D) Programs within the DOE's Clean Coal R&D Program to achieve this mission. A broad portfolio of technologies is being pursued along multiple technology paths to mitigate the risks inherent to R&D. The objective of this report is to use energy systems analysis and conceptual computer simulation models to quantify the impact of this portfolio of technologies on future power generation configurations. This report focuses on bituminous coal feedstock for Integrated Gasification Combined Cycle (IGCC) and Integrated Gasification Fuel Cell (IGFC) power plant configurations that do not capture CO_2. A second volume is underway to provide the same analysis for technologies that will improve the performance and reduce the cost for IGCC and IGFC power plants that employ carbon capture and sequestration.

Approach

The power plant configurations analyzed in this study were modeled using the ASPEN Plus™ modeling program. Emerging technologies were incorporated step-wise over time into the reference IGCC configuration to lay out a "pathway" of technology development and implementation. To the extent possible, a nominal 600 MW plant size was used for comparison between cases. Performance and process limits for advanced technologies were based upon information obtained from the technology developers or published technical reports. Cost estimates for novel technologies were provided by the vendors, or were scaled from existing design/build utility projects and best engineering judgment. Performance and capital and operating costs for conventional equipment were based on the "Cost and Performance Baseline for Fossil Energy Plants, Volume I", DOE/NETL-2007/1281. Capital costs reported are at the total plant cost level and do not include owner's costs, which can be substantial. Care must be taken to avoid comparing the capital costs of this report with those often reported for power plant projects under development, as the latter usually include significant owner's costs. Levelized cost of electricity was determined for all plants assuming investor owned utility financing.

Results

The cumulative impact of the portfolio of advanced technologies in DOE's Clean Coal R&D Program results in power plant configurations that are significantly more efficient and affordable than today's limited set of fossil energy technologies. In the IGCC process alone, there is the potential for 11 percentage point improvement over conventional gasification technology. With fuel cell technology, process efficiency improvements upwards of 24 percentage points are potentially achievable. Capital cost reductions result not only from less expensive technology alternatives such as warm gas cleanup and ITM air separation, but also from increased power generation brought about by advanced technology such as syngas turbines – resulting in cumulative total plant cost reductions by as much as $700/kW after all advanced technologies are implemented. Improvements in process efficiency, reductions in capital and operating expense, and increase in capacity factor all contribute to decreased cost of electricity (COE), projecting an overall decrease by more than 3 cents/kW-hr – or a decrease of 35 percent.

Results of the analysis clearly indicate that the current portfolio is capable of achieving the specific cost and efficiency goals set out by the Clean Coal R&D Program. The results also highlight the importance of continued R&D, large-scale testing, and integrated deployment so that these technologies are proven to the point where they become commercially-accepted technology for future coal-based power plants.

ACKNOWLEDGEMENT

The authors wish to thank Julianne Klara and Gary Stiegel of DOE's National Energy Technology Laboratory (NETL) for helpful discussions and suggestions during the execution of this work. Numerous technical contributions were also provided by NETL's Richard Dennis, John Wimer, and Wayne Surdoval. Dale Keairns and Richard Newby of SAIC were instrumental in providing the pressurized solid oxide fuel cell process design.

This work was funded by the U.S. DOE's National Energy Technology Laboratory and performed by Noblis under a contract with the U.S. Department of the Interior (NBCH-D-02-0039/0051) and Interagency Agreement DE-AI26-04NT4229 1.

EXECUTIVE SUMMARY

The United States Department of Energy's (DOE) Strategic Center for Coal funds research and development (R&D) whose objective is to improve the efficiency and reduce the cost of advanced power systems. In order to evaluate the benefits of on-going R&D, Noblis utilized their energy systems analysis capabilities and conceptual computer simulation models to quantify the impact of successful federally-funded R&D on future power systems configurations.

A variety of process scenarios that produce electric power from bituminous coal are analyzed in this study. Starting with a reference integrated gasification combined cycle (IGCC) plant using conventional technology, a series of process modifications are made to represent commercialization of advanced technologies. Impacts on both process performance and cost are evaluated. Technology development is examined from two perspectives: the first examines the individual contribution of each new advanced technology, and the second examines the cumulative impact as each technology is added to the most advanced process configuration. In this manner, the contribution of DOE's R&D program to future power systems technology can be measured and prioritized.

A focus on non-carbon capture cases represents the first phase, or Volume 1, of this pathway study; the second phase, to be addressed by a follow-on report (Volume 2), will examine processes involving carbon capture.

Volume 1 is organized into two parts. The body of this report presents an executive-level analysis of performance and cost for each case, and expected trends over time. The Supplement to Volume 1 is intended as a reference for engineers involved in systems analysis of processes similar to those conducted in this study. It provides additional detail regarding process flow diagrams, process descriptions, computer modeling approaches, capital equipment costs, economic assumptions, and detailed results comparisons between cases.

Table ES-1. Power System Technology Development

Case	Description
0	Reference Plant / Slurry Feed Gasifier / Cryogenic ASU / Cold Gas Cleanup / 7FA Syngas Turbine / 75 % Capacity Factor (2002 Technology)
1	Slurry Feed Gasifier / Cryogenic ASU / Cold Gas Cleanup / 7FA Syngas Turbine / **80 % Capacity Factor**
2	Slurry Feed Gasifier / Cryogenic ASU / Cold Gas Cleanup / **Advanced "F" Frame Syngas Turbine** / 80 % Capacity Factor
3	**Coal Feed Pump** / Cryogenic ASU / Cold Gas Cleanup / Advanced "F" Frame Syngas Turbine / 80 % Capacity Factor
4	Coal Feed Pump / Cryogenic ASU / Cold Gas Cleanup / Advanced "F" Frame Syngas Turbine / **85 % Capacity Factor**
5	Coal Feed Pump / Cryogenic ASU / **Transport Desulfurizer (TDS) and Direct Sulfur Reduction Process (DSRP)** / Advanced "F" Frame Syngas Turbine / 85 % Capacity Factor
6	Coal Feed Pump / Cryogenic ASU / TDS and DSRP **Warm Gas Treatment for Ammonia and Mercury** / Advanced "F" Frame Syngas Turbine / 85 % Capacity Factor
7	Coal Feed Pump / Cryogenic ASU / Warm Gas Cleanup / **2010-AST Syngas Turbine** / 85 % Capacity Factor
8	Coal Feed Pump / **Ion Transport Membrane (ITM)** / Warm Gas Cleanup / 2010-AST Syngas Turbine / 85 % Capacity Factor
9	Coal Feed Pump / ITM / Warm Gas Cleanup / **2015-AST Syngas Turbine** / 85 % Capacity Factor
10	Coal Feed Pump / ITM / Warm Gas Cleanup / 2015-AST Syngas Turbine / **90 % Capacity Factor**
11	Catalytic Gasifier / Cryogenic ASU / Warm Gas Cleanup / **Pressurized Solid Oxide Fuel Cell** / 90 % Capacity Factor

Reference Case Design Basis

Case 0 defines the reference, or 2002 "vintage" IGCC configuration that uses conventional technology. That process features a single-stage slurry feed gasifier with radiant-only gas cooler followed by Selexol acid gas removal, a 7FA syngas turbine, and conventional three-pressure level steam cycle. Gasifier oxygen is provided by a cryogenic air separation unit (ASU). Process operation assumes a 75 % capacity factor. This IGCC configuration represented conventional technology when DOE established advanced power system R&D program goals in 2003, and is the appropriate baseline against which to evaluate performance and cost improvements resulting from advanced technology.

Process Improvements from Advanced Technologies

The pathway study incorporates new R&D technology into appropriate advanced process configurations in order to examine the cumulative impact of DOE-sponsored technology development over time.

Starting with the reference IGCC configuration, capacity factor increases to 80 % in Case 1 reflecting improvements gained by operating experience from DOE's demonstration program. Case 2 replaces the 7FA syngas turbine with and advanced "F" frame turbine. Case 3 replaces the coal slurry feed to the gasifier with dry feed by incorporating a coal feed pump. Capacity factor increases from 80 % to 85 % in Case 4, reflecting increased process reliability and availability stemming from advanced materials and process control. In Case 5, the cold gas cleanup section is replaced by partial warm gas cleanup: a transport desulfurizer (TDS) and direct sulfur reduction process (DSRP) followed by cold gas ammonia and mercury removal. Case 6 is identical to Case 5, except that the cold gas ammonia and mercury removal process steps are replaced with novel, warm gas treatment processes that result in full warm gas cleanup.

Case 7 represents improvements to the advanced "F" frame turbine by the 2010 timeframe; this advanced syngas turbine is termed the 2010-AST turbine. Case 8 replaces the conventional cryogenic ASU with an ion transport membrane (ITM) to produce oxygen for the gasifier. Case 9 replaces the 2010-AST turbine with an even more advanced syngas turbine (2015-AST) for the 2015 timeframe. In Case 10, the capacity factor again increases – this time from 85 % to 90 % to reflect additional operating experience and improvements in control and materials gained through DOE/NETL's demonstration program.

Finally, Case 11 incorporates a solid oxide fuel cell (SOFC); this case features a catalytic gasifier and has no combustion gas turbine. The process configurations with cumulative process improvements are summarized in Table ES-1 below, with technology changes between cases indicated in bold lettering.

Aspen Plus process simulations were formulated for each plant configuration to compute mass and energy balances and net process efficiencies. Based on plant capacity, conceptual capital cost estimates were developed, and the 20-year levelized cost of electricity was calculated for each case. For consistent comparison, all cost analyses were based on January 2007 dollars, and assumed construction to begin in January, 2007 with a 36-month construction schedule. Costs were based on those developed in NETL's Baseline Study [1], and the same methodology as in the Baseline Study was used to compute levelized cost of electricity. The Aspen Plus simulation and cost estimate from Case 2 were validated against NETL's Baseline Study Case 1, which has an identical process configuration: nearly identical performance and cost results were obtained.

Process simulations and economic evaluations were conducted for all advanced technologies in a stand-alone mode in order to analyze individual contributions from each technology, and also as a form of model validation. In particular, performance predictions for the warm gas cleanup and the ITM cases were validated against the respective technology developers' performance goals. Results from the stand-alone analyses are described in the body of this report, and validation results are provided in the appendix.

Cumulative Impact of Advanced Technologies on Process Efficiency

Figure ES-1 shows the cumulative improvement in process performance as each technology is introduced. Cases that feature improved capacity factor (80 % CF, 85 % CF, and 90 % CF) do not contribute to performance efficiency because the capacity factor merely increases the time of on-stream operation, and therefore has a benefit solely in terms of reduced COE.

The advanced "F" frame turbine and coal feed pump contribute 2.5 and 2.1 percentage point efficiency improvements, respectively. These are slightly greater than the sum of their individual efficiency improvements in the reference plant, so some synergy results from the combined technologies.

Partial warm gas cleanup (WGCU) likewise improves performance of the cumulative process (by 2.1 percentage points) more than it does the performance of the reference plant (2.0 percentage points). Full warm gas cleanup (WGCU+) adds little to performance in the cumulative process (only 0.3 percentage points) because elimination of the ammonia quench, which avoids condensing moisture from fuel gas in the slurry feed gasifier case, does not represent as much of an advantage in a dry feed gasifier whose syngas has virtually no moisture in it.

The 2010-AST turbine, ITM, and 2015-AST turbine cases each improve process efficiency (1.0, 0.9, and 2.0 percentage points, respectively) by a slightly greater amount than they improve the reference plant (0.9, 0.8, and 1.7 percentage points, respectively). This again demonstrates some synergy resulting from combined technologies.

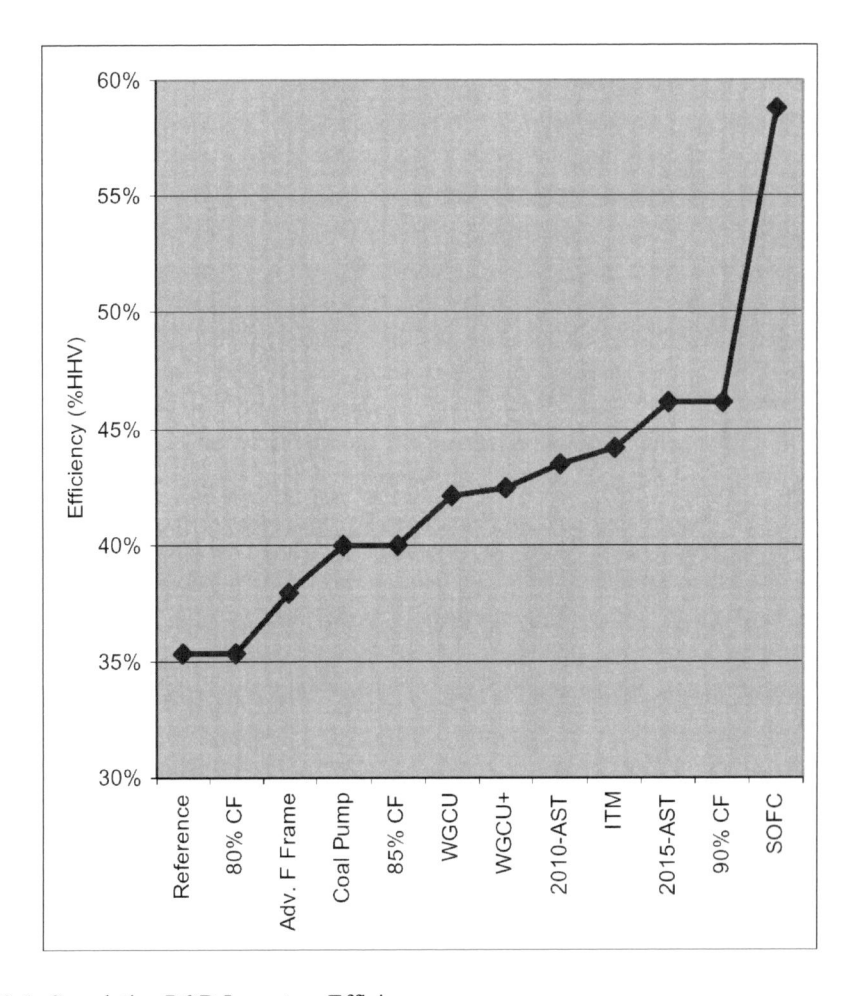

Figure ES-1. Cumulative R&D Impact on Efficiency.

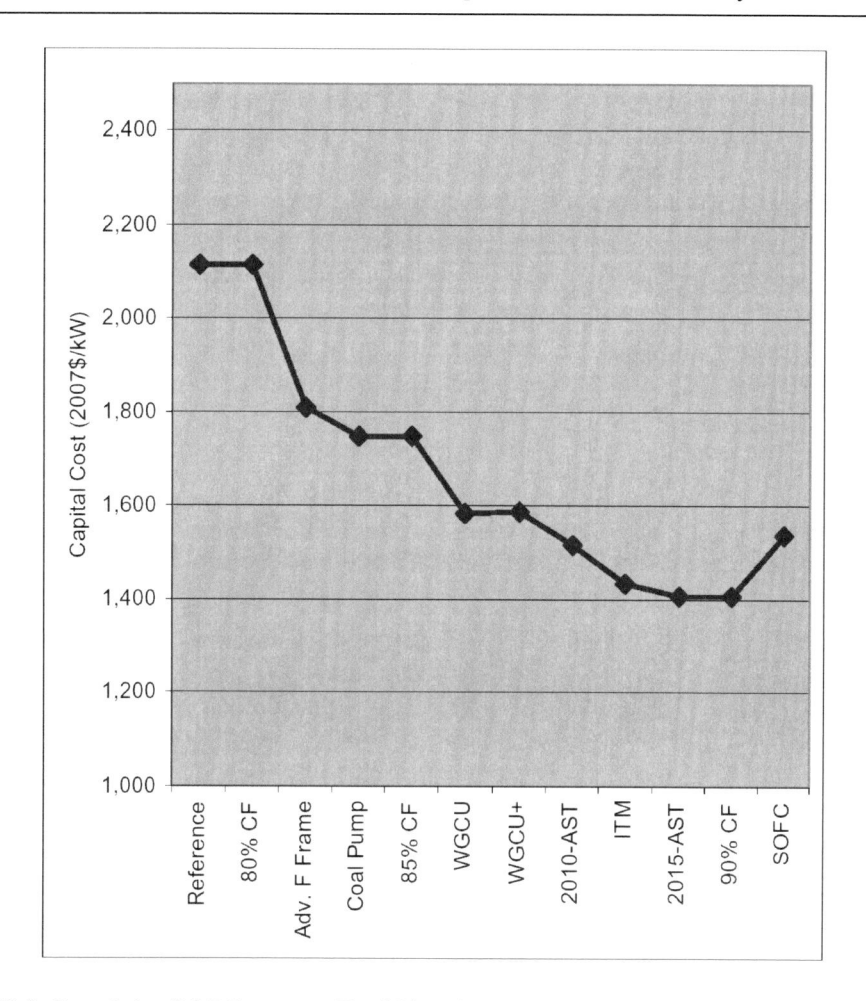

Figure ES-2. Cumulative R&D Impact on Total Plant Cost.

The SOFC process yields 58.8 % plant efficiency. This process relies on a catalytic gasifier with very high (92.0 %) cold gas efficiency and full warm gas cleanup. Compared to the reference process, this represents a very substantial 23.4 percentage point improvement in process efficiency. The high plant efficiency is environmentally attractive because it reduces the production of CO_2 per MWe of power produced. In addition, the process produces a sequestration-ready CO_2 stream, resulting in a superior process from the perspective of cost of CO_2 avoided.

Cumulative Impact of Advanced Technologies on Total Plant Cost

As each advanced technology is introduced, total plant cost usually decreases, as shown in Figure ES-2. Improved capacity factor has no effect on TPC, just as it had no effect on process efficiency.

The advanced "F" frame turbine has the greatest effect of any technology on the cumulative TPC reduction ($304/kW); this is because of the large increase (150 MW) in net power output relative to replacing the 7FA syngas turbine. The incremental reduction from the 2010-AST turbine is $72/kW – not as dramatic a decrease because the power output with the 2010-AST is only 50 MW more than with the advanced "F" frame turbine. The incremental capital cost reduction from the 2015-AST turbine is only $15/kW; this is because of a large decrease (223 MW) in net power output from the plant because the number of trains is reduced from two to one in order to maintain nominal plant output of 600 MW.

As in the reference plant, warm gas cleanup and ITM have lower capital costs than the technologies that they replace, but TPC on a $/kW basis further decreases because net power produced by the plant increases by about 50 MW as a result of each of these technologies. These technologies contribute cumulative technology cost reductions of $1 64/kW and $11 8/kW, respectively. The cost difference between partial warm gas cleanup and full warm gas cleanup is negligible.

For the coal feed pump case, the dry feed gasifier section itself is only slightly less costly than the slurry feed gasifier section (by $24 MM), but the plant as a whole reduces in cost (by $80 MM) due to decreased coal flowrate which results in reduced oxygen requirement and smaller equipment sizes throughout the plant. Accounting for the reduced power output (by 23 MW) of the coal feed pump plant, TPC decreases by $60/kW.

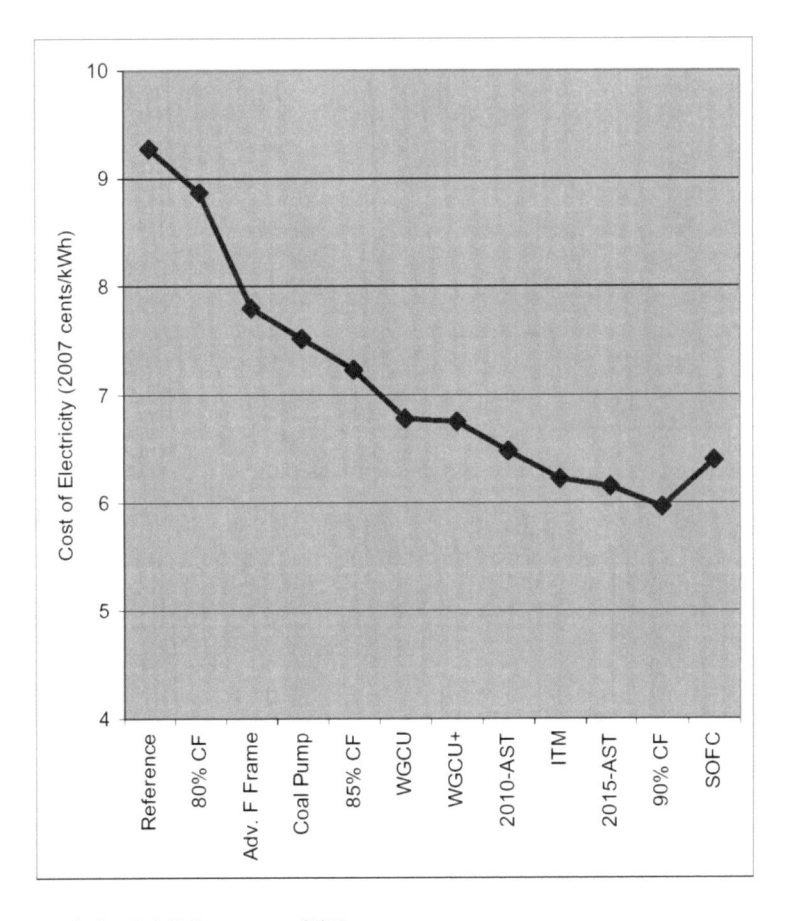

Figure ES-3. Cumulative R&D Impact on COE.

As for the solid oxide fuel cell process, no systems analysis attempt was made to investigate an optimum process configuration; potential for further cost reduction could possibly result from ITM air separation or water gas shift of the fuel gas before it enters the fuel cell or other similar process modifications that may decrease total plant cost. The increase by $153/kW over the most advanced IGCC process with 90 % capacity factor (Case 10) is an artifact of the assumed capital costs of the fuel cell system and catalytic gasifier, and has considerable uncertainty at this time.

Cumulative Impact of Advanced Technologies on COE

As each new advanced technology is implemented step-wise in the cumulative advanced power system, the reduction in COE is represented in Figure ES-3. Due to greater on-stream operation, effects of improved capacity factor are as significant as the other advanced technologies. The increase to 80 % capacity factor results in a 4.0 mills/kW-hr decrease in COE, the increase to 85 % capacity factor results in a 2.9 mills/kW-hr decrease, and the increase to 90 % capacity factor results in a 2.7 mills/kW-hr decrease.

The advanced "F" frame syngas turbine provides the single greatest decrease in COE (10.7 mills/kW-hr) due to the 150 MW increase in net power output and 2.5 percentage point plant efficiency increase made possible by air integration, improved turbine efficiency, and increased HRSG inlet temperature (allowing increased steam cycle superheat and reheat temperatures).

Partial warm gas cleanup results in a 4.5 mills/kW-hr decrease in COE. Because of very low moisture content in the fuel gas, the novel ammonia and mercury removal units in the full warm gas cleanup case result in a very small improvement in process efficiency. There is no significant difference in TPC between partial and full warm gas cleanup, so as a result the COE changes very little between these cases.

The 2010-AST syngas turbine increases plant power output by 50 MW over that of the advanced "F" frame turbine, and therefore results in a $72/kW reduction in TPC. There is also a 1.1 percentage point improvement in process efficiency over the advanced "F" frame turbine, resulting in reduced fuel cost. Overall, the 2010-AST turbine decreases COE by 2.7 mills/kW-hr in the cumulative technologies plant.

The ITM increases plant output by 49 MW with a corresponding decrease in TPC by $17 MM, resulting in a $11 8/kW decrease in total plant cost. Although the efficiency improvement is only 0.9 percentage points, the decreased TPC translates to a 3.6 mills/kW-hr decrease in COE.

The 2015-AST syngas turbine has a much higher power rating than the 2010-AST, but the reduction from two trains to a single train decreases the net plant power output by 223 MW resulting in only a $ 15/kW reduction in TPC and, therefore, a 0.5 mills/kW-hr reduction in COE.

The tremendous process efficiency (58.8 %) and low capital cost ($1,536/kW) of the SOFC process makes its COE competitive with the advanced IGCC processes, even before any systems analysis attempts are made at improved SOFC process configurations. Although the SOFC configuration examined in this study does not have the lowest COE, it represents great potential for carbon capture scenarios because the CO_2 product stream is sequestration-ready.

In summary, this pathway study evaluated anticipated process performance improvements and capital cost reductions resulting from advanced technology development sponsored by DOE. The technology pathway covers a time span of about eighteen (18) years, allowing for the process of technology development and implementation. These advanced technologies include innovations in gasification, syngas turbines, synthesis gas cleaning, air separation, and fuel cells.

The technology improvements examined in this study suggest significant reductions in the COE generated by these advanced power facilities and the value of combining advanced technology to capitalize on their synergistic impacts. Overall, this pathway study determined that DOE/NETL's current R&D portfolio has the potential to reduce total plant cost (TPC) by 35 % and increase efficiency by 24 percentage points by 2020, resulting in a 37 % reduction in cost of electricity (COE).

1. INTRODUCTION

The United States Department of Energy's (DOE) Strategic Center for Coal funds research and development (R&D) whose objective is to improve the efficiency and reduce the cost of advanced Integrated Gasification Combined Cycle (IGCC) and Integrated Gasification Fuel Cell (IGFC) technologies. In order to evaluate the benefits of ongoing R&D, Noblis utilized their energy systems analysis capabilities and conceptual computer simulation models to quantify the impact of successful federally-funded R&D on future power generation configurations.

Noblis developed Aspen Plus computer models for one IGFC and a series of IGCC configurations. These models provided material and energy balances to simulate the gasification of coal to clean synthesis gas and the subsequent utilization in syngas turbine, fuel cell, and steam turbine cycles. Economic models estimated capital and operating costs and calculate the 20-year levelized cost of electricity (COE) based upon standard discounted cash flow (DCF) analysis. An Aspen Plus simulation and cost estimate for one case were validated against a corresponding NETL Baseline Study [1] case, and were found to predict nearly identical performance and cost results.

Emerging advanced gasification, gas cleanup, air separation, syngas turbine, and solid oxide fuel cell technologies were incorporated step-wise over time into the reference IGCC configuration to lay out a "pathway" of technology development and implementation. Incorporation of these advanced technologies into the composite IGCC plant allows an estimate of the future benefits of these technologies to be quantified. These benefits are measured ultimately in terms of reduced cost of electric power.

In this report, a Reference Case IGCC configuration was established based on 2002 technology. Sequential improvements were then evaluated over the anticipated timeframe of advanced technology deployment. These improvements included advanced "F" frame syngas turbine, coal feed pump, greater on-stream time (or capacity factor), warm gas cleanup, improved advanced syngas turbine (2010-AST turbine), ceramic membrane technology for air separation, a further advanced syngas turbine (2015-AST turbine), and emergence of the pressurized solid oxide fuel cell. Increased capacity factor was attributed to advances in instrumentation and materials as well as operating experience gained from demonstrating these technologies over time through DOE programs including Clean Coal Technology, the

Clean Coal Power Initiative, and FutureGen. To the extent possible, a nominal 600 MW plant size was used for comparison between cases.

2. PATHWAY STUDY BASIS

A process flow diagram of the Reference Case is provided in Figure 2-1. This configuration was considered to be state-of-the-art when goals for the advanced power systems program were established in 2003, and is the standard against which all improvements are measured in this study. The process includes two 7FA gas turbines and a steam cycle operating at 1,800 psig with 1,000 °F steam superheat and 1,000 °F steam reheat. The as-received Illinois #6 bituminous coal feed contains 11.12% moisture, and has a higher heating value of 13,125 Btu/lb (dry basis).

A cryogenic air separation unit (ASU) provides oxygen for the single-stage, slurry feed, oxygen- blown gasifier. The ASU is sized to provide sufficient oxygen to the gasifier, plus a small slipstream of oxygen used in the Claus furnace for acid gas treatment. Most of the N2 by-product can be compressed and injected into the topping combustor of the gas turbine; the exact amount is determined by the gas turbine power rating, which is regulated to 192 MW per unit.

Although the gasifier exceeds 2,400 °F during operation, the radiant gas cooler reduces exit raw gas temperature to 1,250 °F. The capacity of a single gasifier is on the order of 2,200 tons/day coal.

Exiting the gasifier, raw fuel gas is scrubbed with water to remove particulates. Water is separated from the slag, and flows to the sour water stripper for treatment. Raw fuel gas is cooled to 390 °F for COS hydrolysis. Following the exothermic COS hydrolysis reaction, the gas is cooled again; first to 310 °F to recover useful heat for fuel gas reheat and steam generation, next to 235 °F to recover useful heat for the steam cycle deaerator, then finally to 110 °F for NH3 removal. The cooling temperatures of 310 °F and 235 °F were selected based on reasonable temperature approaches to the steam cycle streams.

The fuel gas enters packed carbon bed absorbers to remove mercury, followed by a Selexol process that absorbs H2S from the fuel gas. H2S is stripped from the solvent in the solvent regenerator and the acid gas is sent to the Claus plant.

The Claus plant converts H2S to elemental sulfur through a series of reactions. Sulfur is condensed, and tail gas is hydrogenated to convert residual SO_2 back into H2S, which can be captured when the tail gas is recycled to the Selexol absorber. A small slipstream of clean fuel gas is used for reactant.

Clean fuel gas exits the Selexol absorber at 719 psia, and is delivered to the topping combustor at 464.7 psia. Therefore, it can be expanded to recover excess pressure prior to entering the topping combustor; this expansion results in about 6 MWe of power generation.

Fuel gas is diluted with N2 from the ASU. The syngas mixture is burned in the topping combustor, reaching a temperature of 2,250 °F (fuel flow is regulated in order to obtain this temperature). The net gas turbine power output is 192 MWe per unit [2].

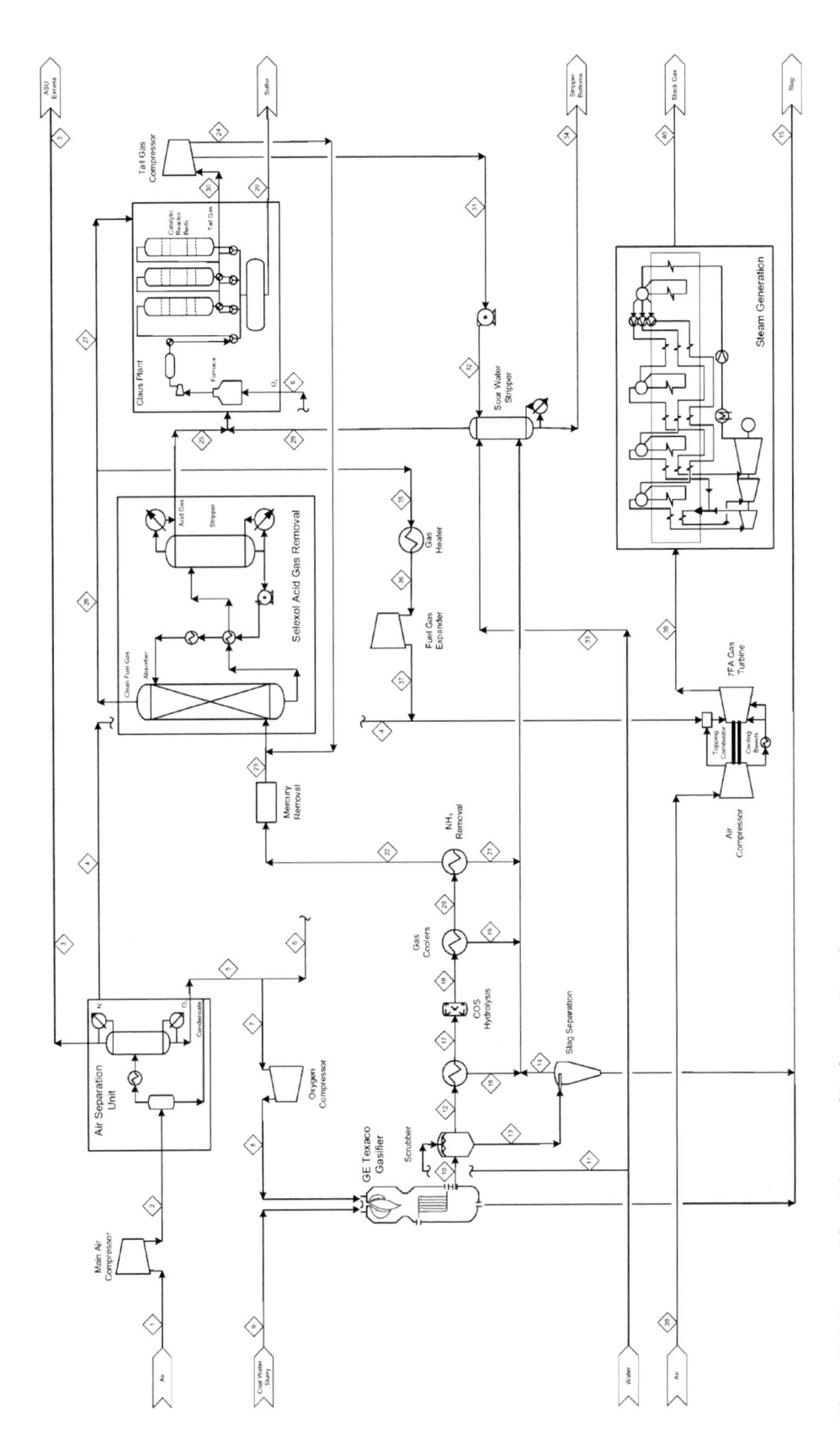

Figure 2-1. Process Flow Diagram of Reference Case 0.

Table 2-1. Coal Analysis: Illinois #6 Old Ben #26 Mine

Proximate Analysis As-Received (wt %)	
Moisture	12.51
Ash	10.91
Volatile Matter	39.37
Fixed Carbon	49.72
Ultimate Analysis Dry Basis (wt %)	
Ash	10.91
Carbon	71.72
Hydrogen	5.06
Nitrogen	1.41
Chlorine	0.33
Sulfur	2.82
Oxygen	7.75
Total	100.00
HHV (Btu/lb)	13,126

All available process heat is collected for steam generation in the bottoming cycle. Superheated steam is expanded through three turbines, with reheat after the high pressure turbine. The steam cycle also provides heat for acid gas removal (the Selexol solvent regenerator), the sour water stripper, and fuel gas reheating prior to the fuel gas expander.

The design basis of NETL's Baseline Study was adopted so that results from this pathway study would be consistent with established results. Some of the more global process parameters are described below, while other case-specific design assumptions can be found in the Volume 1 Supplement along with more detailed documentation for each individual case.

2.1. Coal Analysis

Fuel quality has a significant effect on process performance. Table 2-1 details the coal feed analysis that is used for all cases in this study. The Illinois #6 bituminous coal comes from the Old Ben #26 mine, and is the same as used in NETL's Baseline Study. Note the fuel heating value of 13,126 Btu/lb (dry basis); this is equivalent to 11, 666 Btu/lb for as-received coal.

This coal has a relatively high chlorine content, which will be shown in the analysis to impact the sour water stripper operation in order to prevent corrosion; a water purge stream maintains chloride concentration below 1,000 ppm in the sour water stripper.

2.2. Process Operating Assumptions

The cryogenic ASU operates at 10 atmospheres, producing 95 % pure oxygen for the gasifier (and Claus plant if cold gas cleanup is used). Nitrogen is used to dilute fuel to the gas turbine; in most cases, nitrogen is added to regulate fuel gas heating value to 125 Btu/scf

(LHV) as a method for NOx control. If there is not sufficient nitrogen for dilution, steam is added to the fuel stream to meet the fuel specification. The ASU consumes a small quantity of low pressure steam to regenerate dehumidification sorbent, and a small quantity of medium pressure steam for an ammonia refrigeration system.

The slurry feed gasifier is assumed to operate at 2,400 °F with 98 % carbon conversion; the gasifier with dry feed provided by the coal feed pump is assumed to operate at 2,600 °F with 99.5 % carbon conversion. Gasifier pressure is 800 psia. In both cases, sufficient oxygen is provided to the gasifier to satisfy the energy balance. Exiting the gasifier, the radiant-only cooler reduces the raw syngas stream temperature to 1,250 °F. No convective cooler is present in cold gas cleanup cases, however a trim cooler is used to control temperature of the transport desulfurizer in the warm gas cleanup cases.

In cold gas cleanup processes, raw fuel gas exiting the gasifier is scrubbed with water to remove particulates. Water is separated from the slag, and flows to the sour water stripper for treatment. Raw fuel gas is cooled to 390 °F for COS hydrolysis. Following the exothermic COS hydrolysis reaction, the syngas is further cooled to condense water, ammonia, and cyanide and also to prepare for H2S removal in the Selexol absorber. During this cooling, heat is assumed to be recoverable down to 235 °F for use in the bottoming cycle.

Condensate from the raw syngas is treated in the sour water stripper; a water purge is added to the sour water stripper in order to maintain chloride concentration below 1,000 ppm. Gas component separation in the Selexol process is based on proprietary information provided by UOP. Acid gas (stripped from the regenerated Selexol solvent) is treated in a Claus plant for sulfur recovery, and the tail gas is recycled to the Selexol absorber.

Four syngas turbines are examined in this pathway study; the 7FA, advanced "F" frame, 2010-AST, and 2015-AST. These turbines are all designed for operation using syngas fuel. They operate at increasing firing temperatures, pressure ratios, and power ratings as technology improves over time. All turbines except the 7FA are integrated with the ASU – providing part of the ASU's air feed in order to reduce the work required of the main air compressor. The 7FA and advanced "F" gas turbines are now commercially available; the 2010-AST and 2015-AST (pseudonyms are used for the purpose of discussion) represent technology expected to be available in the 2010 and 2015 timeframes, respectively.

Turbine exhaust flows to the heat recovery steam generator (HRSG), which provides heat for a three pressure level steam cycle. High pressure steam superheat and reheat is 1,000 °F for the 7FA syngas turbine, but increases to 1,050 °F as turbine firing temperatures (and exit temperatures) increase in the other three turbine models. For all cases, the flue gas stack exit temperature is 270 °F; standardization of this value provides consistency to the quantity of recoverable heat from the HRSG.

2.3. Economic Analysis

The cost estimating methodology used in this study is consistent with that described in NETL's Quality Guidelines for Energy Systems Studies (QGESS) [3]. Plant capital cost is estimated using the most accurate estimation methods available, taking into consideration plant size, number of process trains, sparing philosophy, and as much equipment-specific design information as possible. In general, scaling factors are used to calculate equipment costs based on capacity or throughput; the Supplement to Volume 1 describes specific methods used to estimate costs of the advanced technologies.

Operating and maintenance (O&M) costs include fixed labor costs as well as variable costs (that depend on capacity factor) including maintenance materials, water, chemicals, and waste disposal. Fuel cost is calculated separately from O&M.

Economic feasibility analysis can be performed using the Power Systems Financial Model (PSFM) [4], Version 5.0. Alternatively, the cost of electricity calculation (described below) can be based directly on the capital charge factor. This study assumes a prescribed capital charge factor (17.5 %) typical of a higher-risk project undertaken by an investor-owned utility [5].

2.3.1. Capital Cost

The following Figure 2-2, taken from Chapter 6 of QGESS, illustrates the relationships between various elements of capital cost. Noblis correlations are used to estimate Bare Erected Cost (BEC) for each major section of the process plant. The BEC is estimated (in January 2007 dollars) using mass and energy balance information from Aspen Plus simulations of each case. For ease in comparing results, the organization of plant sections is consistent with the presentation used in NETL's Baseline Study. Each section's BEC represents the sum of major plant equipment within the section (including initial chemical and catalyst loadings), as well as materials and labor. Appropriate for a scoping study, BEC's are based on scaled estimates using best-available information collected from multiple sources for the cost correlations.

The BEC is used as the basis for calculating detailed engineering and construction and project management fees. A 9 % charge is applied which, when added to the BEC, becomes the Engineering, Procurement, and Construction Cost (EPCC). The cost analyses in Chapter 3 of this report present the EPCC at the process section level; however the Volume 1 Supplement contains additional process section detail for BEC, EPCC, and process and project contingencies for all cases.

For consistency, process and project contingencies used in NETL's Baseline Study form the basis for all major equipment in each plant section. Advanced technologies are assumed to have the same level of contingency as conventional technologies in order not to put the advanced technologies at a disadvantage due to uncertainties in their cost. Contingency estimates are added to the EPCC to calculate the Total Plant Cost (TPC).

Startup costs (assumed to be 2 % of EPCC), owner's costs (which might typically include a Technology Fee or licensing fee), and the time value of money are normally added to the TPC in order to obtain the Total Required Capital (TRC). For consistency with NETL's Baseline Study, owner's costs are omitted in this economic analysis because they are project-specific. Therefore, the reader should bear in mind that the financial results of this analysis (levelized cost of electricity and capital charge factor) do not include owner's costs.

2.3.2. O&M Cost

Labor represents a fixed operating cost, and is based on the number of operating laborers in the plant. The Baseline Study estimate for number of laborers, labor rates, burden, and administrative overhead is used as a basis. Administrative labor is estimated as an overhead rate (25 %) to the sum of operating and maintenance labor. An average labor rate of $33/hr is assumed – again consistent with that used in NETL's Baseline Study.

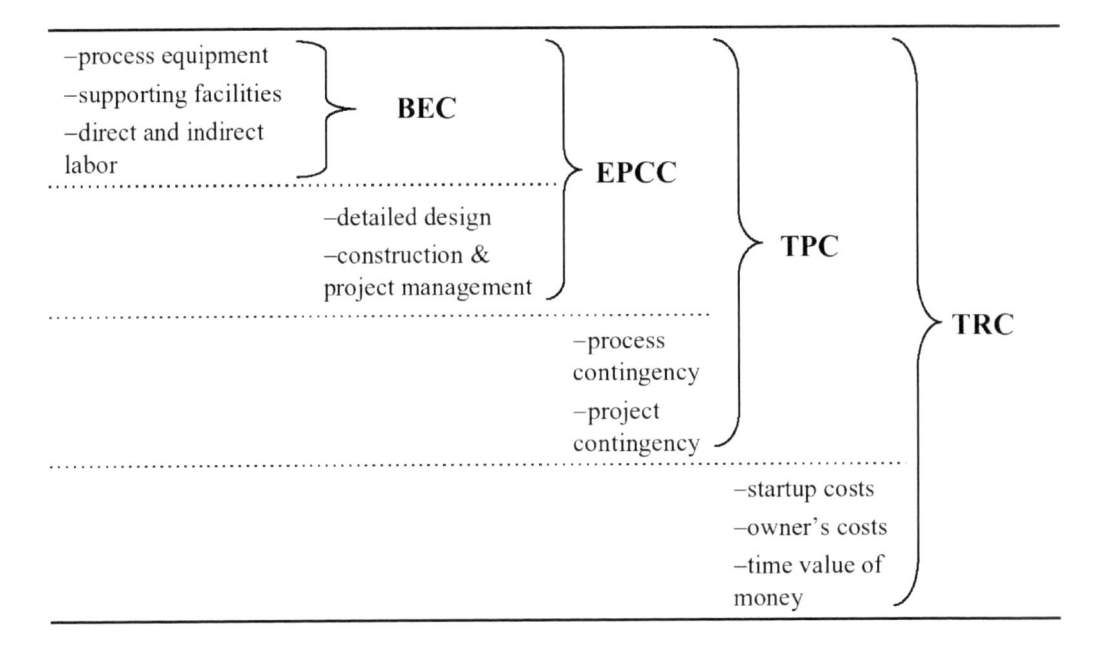

Figure 2-2. Elements of Capital Cost.

Table 2-2. Elements of Variable Operating Cost

Maintenance Materials
Water
Chemicals
Carbon (Hg removal)
COS Catalyst
Shift Catalyst
Claus Catalyst
Selexol Solvent
ZnO Sorbent
Fuel Cell Stack Replacement
Spent Catalyst Waste Disposal
Ash Disposal

Variable operating costs are estimated using 100 % capacity factor, and expressed as percent of EPCC if using the PSFM[1]. The PSFM applies the capacity factor to calculate actual annual variable operating cost. Table 2-2 identifies elements of variable operating cost that are included in the analysis. Consistent with the Baseline Study, no credit is taken for by-products from any process.

The PSFM computes fuel cost based on net power generation, heat rate, and fuel heating value. A coal cost of $42.1 1/ton ($1.80/MMBtu) is assumed, with an as-received heating value of 11,666 Btu/lb. For warm gas cleanup, costs of $14,000/ton for ZnO sorbent and $100/ton for trona are assumed[2]. The sorbent attrition rate is assumed to be 10-20 lb. per million lb. circulating sorbent.

2.3.3. Cost of Electricity

As an alternative to the PSFM, the levelized cost of electricity can be calculated directly using the formula:

$$COE_P = ((CCFP*TPC)+LF_{FP}*FYC_F+CF*(LF_{1P}*FYC_1+LF_{2P}*FYC_2+...))/(CF*MWh)$$

Where:

COE_P = levelized cost of electricity over P years
CCF_P = capital charge factor levelized over P years
TPC = total plant cost
LF_{FP} = levelization factor over P years for fixed operating costs
FYC_F = first year fixed operating costs
CF = capacity factor
LF_{nP} = levelization factor over P years for category n variable operating cost element
FYC_n = first year variable operating costs for category n cost element
MWh = net annual power generation at 100% capacity factor

The capital charge factor can be considered to be the rate at which capital costs are recovered during the lifetime of the project. It is a function of cost of capital and level of technology risk; as these factors increase, the capital charge factor also increases. For the purposes of this study, the investment scenario is considered to be an investor-owned utility (IOU) involved in higher- risk technology. Based on guidance from QGESS, the capital charge factor in this scenario is 17.5 %. Additional assumed financial parameters (used in NETL's Baseline Study) are itemized in Table 2-3 below.

Table 2-3. Discounted Cash Flow Analysis Parameters

Parameter	Value
Percentage Debt	45 %
Interest Rate	11.55 %
Repayment Term of Debt	15 years
Grace Period on Debt Repayment	0 years
Debt Reserve Fund	None
Depreciation	20 years 150 % DB
Working Capital	Zero
Plant Economic Life	30 years
Coal Escalation Factor	2.35 %
O&M Escalation Factors	1.87 %
EPC Escalation	0 %
Tax Holiday	0 years
Income Tax Rate	38 %
Investment Tax Credit	0 %
Duration of Construction	36 months

Individual levelization factors for the COE equation above can be calculated by:

$$LF_{nP} = k * (1-k^P) / (a_P * (1-k))$$

Where

k = $(1+e) / (1 + i)$
a_P = $(((1+i)^P - 1) / (i * (1+i)^P)$
e = annual escalation rate
i = annual discount rate

Consistent with NETL's Baseline Study, the 20-year O&M levelization factors for both fixed and variable costs are 1.1568 (presumes an escalation rate of 1.87 %). For coal, the 20-year levelization factor is 1.2022 (presumes an escalation rate of 2.35 %). Once again, all costs in this analysis are based on January 2007 dollars.

3. ANALYSIS OF ADVANCED POWER PROCESS CONFIGURATIONS

A variety of process scenarios that produce electric power from bituminous coal are analyzed in this study to determine the potential performance improvements and cost reductions resulting from advanced technology under development in DOE/NETL's Clean Coal R&D program. Starting with the reference IGCC plant, a series of process modifications is simulated to represent commercialization of advanced technologies. Impacts on both process performance and cost are evaluated. The impact of each individual technology is first evaluated within the framework of the reference plant. These process configurations are listed in Table 3-1, with each of the advanced technologies identified in bold letters. The suffix "a" on the case number indicates each technology evaluated singly in the reference plant.

Table 3-1. Stand-Alone Power System Technology Development

Case	Description
0	Reference Plant / Slurry Feed Gasifier / Cryogenic ASU / Cold Gas Cleanup / 7FA Syngas Turbine / 75 % Capacity Factor (2002 Technology)
2a	Slurry Feed Gasifier / Cryogenic ASU / Cold Gas Cleanup / **Advanced "F" Frame Syngas Turbine** / 75 % Capacity Factor
3a	**Coal Feed Pump** / Cryogenic ASU / Cold Gas Cleanup / 7FA Syngas Turbine / 75 % Capacity Factor
5a	Slurry Feed Gasifier / Cryogenic ASU / **Transport Desulfurizer (TDS) and Direct Sulfur Recovery Process (DSRP)** / 7FA Syngas Turbine / 75 % Capacity Factor
6a	Slurry Feed Gasifier / Cryogenic ASU / TDS and DSRP with **Warm Gas Treatment for Ammonia and Mercury** / 7FA Syngas Turbine / 75 % Capacity Factor
7a	Slurry Feed Gasifier / Cryogenic ASU / Cold Gas Cleanup / **2010-AST Syngas Turbine** / 75 % Capacity Factor
8a	Slurry Feed Gasifier / **Ion Transport Membrane (ITM)** / Cold Gas Cleanup / Advanced "F" Frame Syngas Turbine / 75 % Capacity Factor
9a	Slurry Feed Gasifier / Cryogenic ASU / Cold Gas Cleanup / **2015-AST Syngas Turbine** / 75 % Capacity Factor

Table 3-2. Cumulative Power System Technology Development

Case	Description
0	Reference Plant / Slurry Feed Gasifier / Cryogenic ASU / Cold Gas Cleanup / 7FA Syngas Turbine / 75 % Capacity Factor (2002 Technology)
1	Slurry Feed Gasifier / Cryogenic ASU / Cold Gas Cleanup / 7FA Syngas Turbine / **80 % Capacity Factor**
2	Slurry Feed Gasifier / Cryogenic ASU / Cold Gas Cleanup / **Advanced "F" Frame Syngas Turbine** / 80 % Capacity Factor
3	**Coal Feed Pump** / Cryogenic ASU / Cold Gas Cleanup / Advanced "F" Frame Syngas Turbine / 80 % Capacity Factor
4	Coal Feed Pump / Cryogenic ASU / Cold Gas Cleanup / Advanced "F" Frame Syngas Turbine / **85 % Capacity Factor**
5	Coal Feed Pump / Cryogenic ASU / **Transport Desulfurizer (TDS) and Direct Sulfur Recovery Process (DSRP)** / Advanced "F" Frame Syngas Turbine / 85 % Capacity Factor
6	Coal Feed Pump / Cryogenic ASU / TDS and DSRP / **Warm Gas Treatment for Ammonia and Mercury** / Advanced "F" Frame Syngas Turbine / 85 % Capacity Factor
7	Coal Feed Pump / Cryogenic ASU / Warm Gas Cleanup / **2010-AST Syngas Turbine** / 85 % Capacity Factor
8	Coal Feed Pump / **Ion Transport Membrane (ITM)** / Warm Gas Cleanup / 2010-AST Syngas Turbine / 85 % Capacity Factor
9	Coal Feed Pump / ITM / Warm Gas Cleanup / **2015-AST Syngas Turbine** / 85 % Capacity Factor
10	Coal Feed Pump / ITM / Warm Gas Cleanup / 2015-AST Syngas Turbine / **90 % Capacity Factor**
11	Catalytic Gasifier / Cryogenic ASU / Warm Gas Cleanup / **Pressurized Solid Oxide Fuel Cell** / 90 % Capacity Factor

The cumulative impact of all technologies available at any one time is also evaluated. That is, as each new technology becomes available, it is implemented in the composite process to evaluate potential improvements in either process performance or cost over time. Table 3-2 identifies the process configurations of these cases.

3.1. Case 0: Reference Plant

The reference plant is an IGCC process that includes slurry feed gasifier, cryogenic air separation, cold gas cleanup, 7FA syngas turbine, and 75 percent capacity factor. The process configuration is based on state-of-the-art technology available in 2002, and serves as an appropriate metric to evaluate technology progress because it was the basis used in 2003 to establish DOE's R&D program goals.

Figure 3-1 presents a block flow diagram of the process. The plant is configured with two trains of single-stage slurry feed gasifiers with radiant-only syngas coolers, two cryogenic air separation units, two trains of water scrub and carbonyl sulfide (COS) hydrolysis, a single train of Selexol acid gas removal, one train of sulfur recovery using conventional Claus technology, two trains of 7FA syngas turbines, one HRSG, and one steam turbine bottoming cycle with high, intermediate, and low pressure (condensing) turbine sections. Steam conditions are 1,800 psi and 1,000 °F for the HP turbine and 405 psi and 1,000 °F for the IP turbine.

This two-train IGCC plant processes 4,831 tons per day of as-received Illinois #6 coal to produce a net 487 MW of power. Carbon utilization is 98 percent, and overall efficiency is 35.4 percent (HHV basis). Total power generated includes 384 MW from the gas turbines, 6 MW from the fuel gas expanders, and 223 MW from the steam cycle. Auxiliary power use is estimated to be 127 MW. This performance, calculated by Noblis' Aspen Plus process model, is comparable to operation achieved at the Tampa Electric Plant, which uses the same technology.

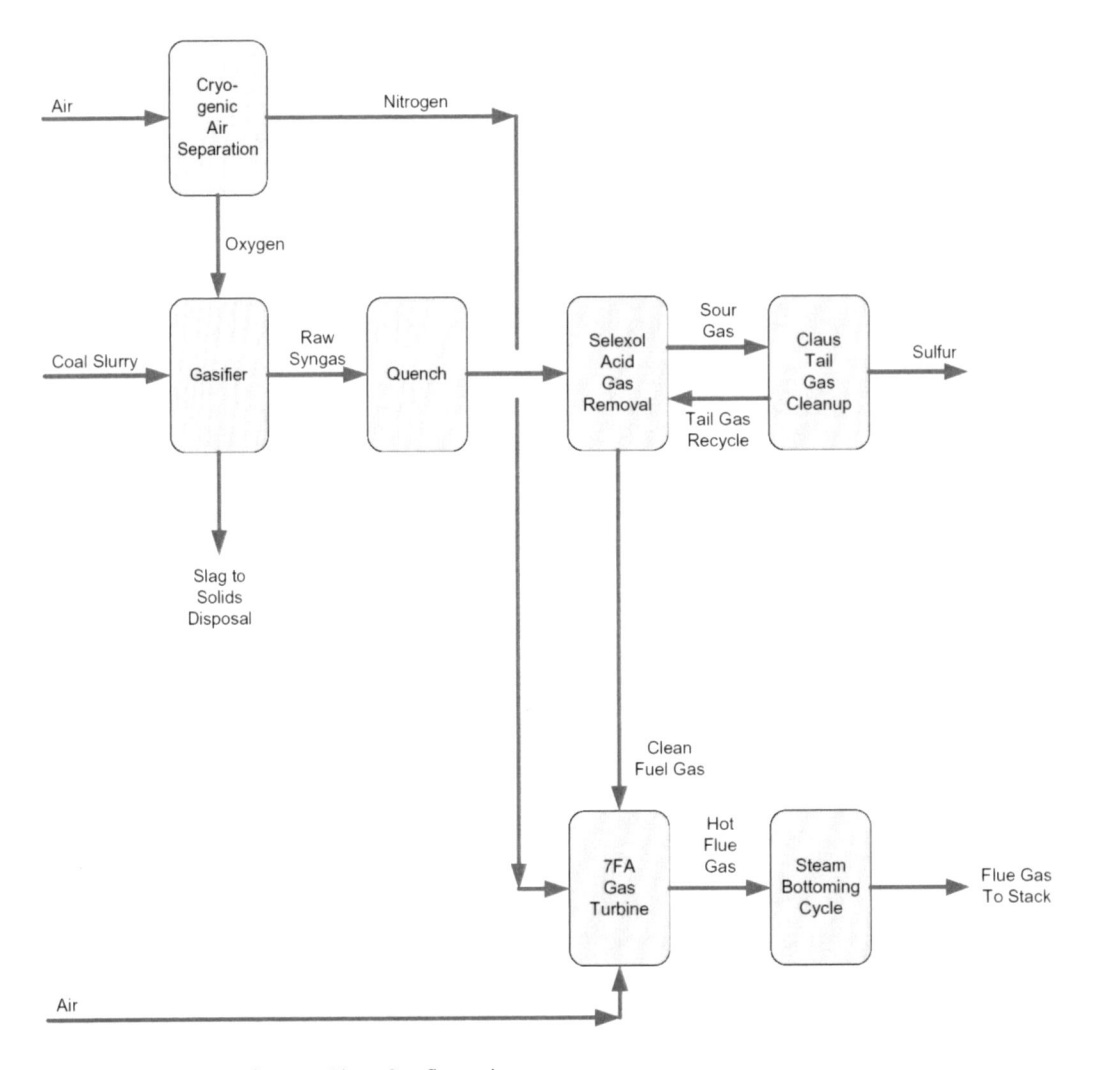

Figure 3-1. Case 0: Reference Plant Configuration.

Cost Analysis

Table 3-3 below estimates the Engineering, Procurement, and Construction Cost (EPCC) for each major section of the process plant. Bare erected costs (BEC) are scaled from equipment costs in NETL's Baseline Study. Process and project contingencies (also from NETL's Baseline Study) are added to the EPCC to calculate the Total Plant Cost (TPC). The TPC does not include owner's costs, which might typically include a Technology Fee. The resulting TPC is $2,1 13/kW.

Table 3-3. Case 0: Capital and O&M Cost Summary

Capital Cost ($1,000)					
Plant Sections	**EPCC**	**Process Cont'gncy**	**Project Cont'gncy**	**TPC**	**TPC $/kW**
1 Coal Handling	25,685	0	5,137	30,821	63
2 Coal Prep & Feed	39,472	1,312	8,195	48,980	101
3 Feedwater & Balance of Plant	28,606	0	6,471	35,077	72
4a Gasifier	184,371	18,725	33,116	236,212	485
4b Air Separation Unit	153,591	0	15,359	168,950	347
5a Gas Cleanup	93,441	75	18,873	112,389	231
5b CO_2 Removal & Compression	0	0	0	0	0
6 Gas Turbine	91,110	3,787	10,161	105,058	215
7 HRSG	44,560	0	4,951	49,511	102
8 Steam Cycle and Turbines	47,842	0	6,467	54,310	112
9 Cooling Water System	20,099	0	4,134	24.233	50
10 Waste Solids Handling System	34,981	0	3,771	38,752	80
11 Accessory Electric Plant	55,772	0	10,757	66,529	137
12 Instrumentation & Control	18,982	869	3,327	23,178	48
13 Site Preparation	13,956	0	4,187	18,143	37
14 Buildings and Structures	14,012	0	2,302	16,314	34
Total	866,482	24,769	137,207	1,028,457	2,113
O&M Cost ($1,000)					
Fixed Costs				**Total**	**% EPCC**
Labor				19,542	2.26
Variable Operating Costs*				**Total**	**% EPCC**
Maintenance Materials				18,368	2.12
Water				1,451	0.17
Chemicals				1,021	0.12
Waste Disposal				2,262	0.26
Total Variable Costs				23,102	2.67
Total O&M Cost*				42,644	4.92
Fuel Cost*				55,690	6.43
Discounted Cash Flow Results					
Total Plant Cost ($/kW)					2,113
Levelized Cost of Electricity ($/kW-hr)					0.0927

*Includes 75 % Capacity Factor

Labor represents a fixed operating cost, and is based on the number of operating laborers in the plant. The Baseline Study estimate for number of laborers, labor rates, burden, and administrative overhead was used for consistency. Administrative labor is estimated as an overhead rate (25 %) to the sum of operating and maintenance labor.

Variable operating costs are estimated using 75 % capacity factor – typical of the availability of IGCC plants in 2002/2003. Levelized cost of electricity is calculated using the equation of the previous section. Results from the discounted cash flow analysis, shown Table 3-3, indicate $0.0927/kW-hr 20-year levelized cost of electricity based on January 2007 dollars.

3.2. Increased Capacity Factor to 80 Percent

With IGCC operating experience gained from DOE's demonstration programs, plant availability is expected to improve to 80 % even without the need for improved technology. Capacity factor has no effect on the mass and energy balance computed in Case 0; only the variable operating costs and fuel cost are affected. Substituting the higher capacity factor into the equation for levelized cost of electricity, the levelized COE improves from $0.0927/kW-hr in Case 0 to $0.0887/kW-hr in Case 1 as the result of more hours of plant operation.

3.3. Advanced "F" Frame Syngas Turbine

The advanced "F" frame syngas turbine allows integration with the air separation unit (a portion of the air supply to the ASU is provided by the gas turbine). The advanced "F" frame syngas turbine produces more power, has a higher pressure ratio, and higher firing temperature than the 7FA syngas turbine. Because of the higher turbine firing temperature and subsequently higher turbine exhaust temperature, steam conditions are 1,800 psi and 1,050 °F for the HP turbine and 405 psi and 1,050 °F for the IP turbine.

3.3.1. Impact of Advanced "F" Frame Syngas Turbine in the Reference Plant (Case 2a)

Case 2a Configuration: Slurry Feed Gasifier, Cryogenic ASU, Cold Gas Cleanup, Advanced "F" Frame Syngas Turbine, 75 % Capacity Factor

Figure 3-2 presents the block flow diagram of the reference IGCC process with an advanced "F" frame syngas turbine. This two-train IGCC plant processes 5,900 tons per day of as-received coal to produce a net 637 MW of power. Overall efficiency is 37.9 percent (HHV basis). Carbon utilization is 98 percent and the capacity factor is 75 percent. Total power generated includes 8 MW from the fuel gas expander, 464 MW from the gas turbines and 293 MW from the steam turbine. Auxiliary power use is estimated to be 128 MW. Performance resulting from the advanced "F" frame gas turbine is compared against the Reference Case in the following table.

Table 3-4. Performance Impact of Advanced "F" Turbine in the Reference Plant

	Case 0	Case 2a
	Reference plant with 7FA	Reference plant with adv. "F"
Gas Turbine Power (MWe)	384	464
Fuel Gas Expander (MWe)	6	8
Steam Turbine Power (MWe)	223	293
Total Power Produced (MWe)	614	765
Auxiliary Power Use (MWe)	-127	-128
Net Power (MWe)	487	637
As-Received Coal Feed (lb/hr)	402,581	491,633
Net Heat Rate (Btu/kW-hr)	9,649	9,004
Net Plant Efficiency (HHV)	35.4 %	37.9 %

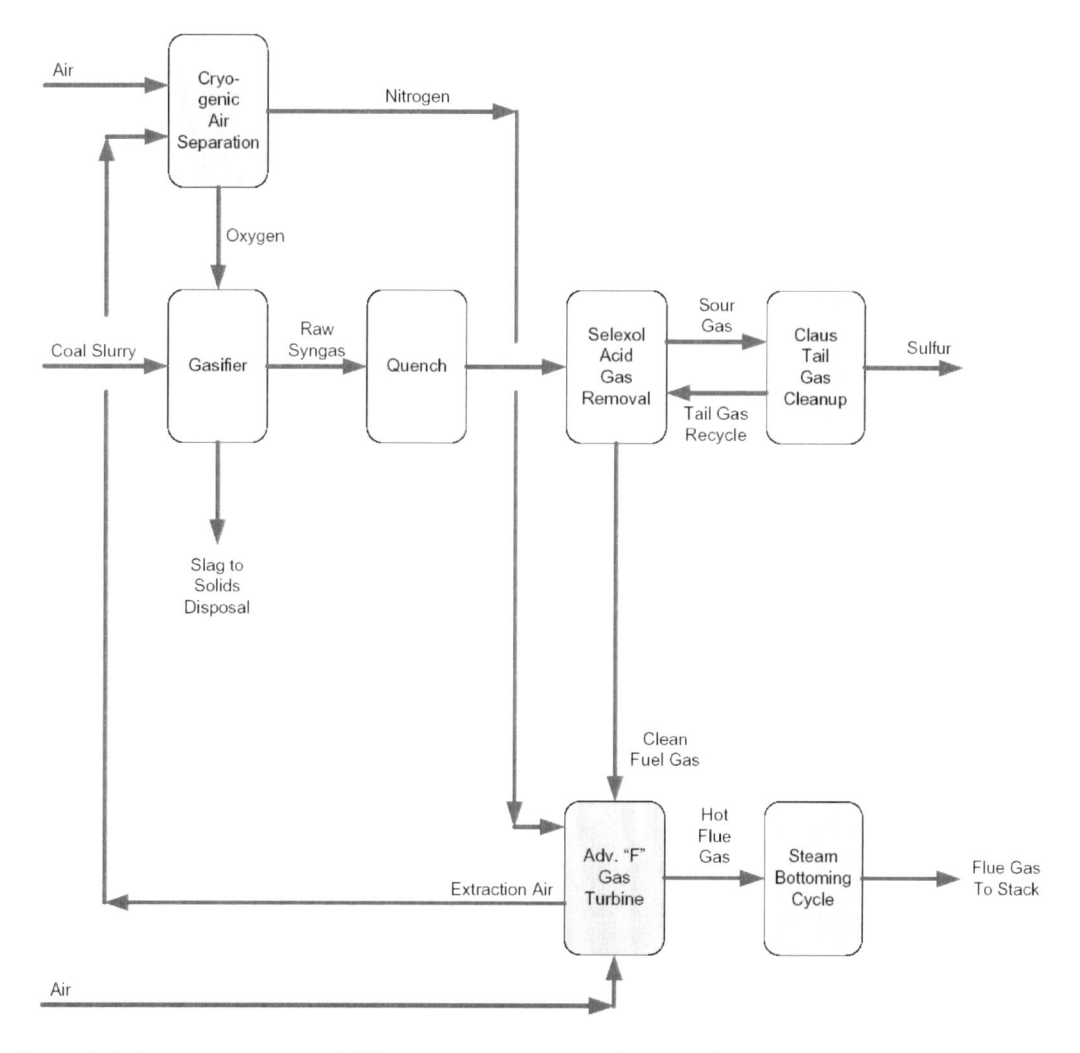

Figure 3-2. Case 2a: Advanced "F" Frame Syngas Turbine IGCC Configuration.

The 7FA syngas turbine in the Reference Case is rated at 192 MW, while the advanced "F" frame turbine in Case 2a is rated at 232 MW. Because of the lower turbine exit temperature in Case 0, steam superheat temperature is 1,000 °F rather than the 1,050 °F that's possible in Case 2a due to the higher turbine exit temperature. The increased coal flowrate, made possible by greater turbine throughput, leads to increased heat recovery in the gasifier, syngas quench, and flue gas through the HRSG – thus further contributing to increased steam turbine power generation in Case 2a.

Although auxiliary power use appears to be nearly the same between cases, there are significant but off-setting differences in the ASU main air compressor and the nitrogen compressor. The ASU main air compressor power consumption decreases in Case 2a due to integration between the gas turbine air compressor and the ASU, which reduces the fresh air feed through the main air compressor and therefore reduces power consumption. This reduction in power consumption is counterbalanced by increased N2 compressor power consumption, which is the result of greater flowrate through the gas turbine. As a fraction of total power produced, auxiliary power use decreases for the larger gas turbine with air integration.

Overall, the net plant efficiency increases by 2.5 percentage points going from the 7FA syngas turbine to the advanced "F" frame syngas turbine; the primary reasons for this are air integration from the gas turbine to the ASU, the higher efficiency of the advanced "F" frame syngas turbine, and the increased steam cycle superheat temperature.

Cost Analysis (Case 2a)

Table 3-5 below compares capital and O&M costs with the Reference Case. The choice of gas turbine is the reason for differences in capital costs between Case 0 (7FA turbine) and Case 2a (advanced "F" frame turbine). The advanced "F" turbine has a higher power rating, which increases coal flowrate to the process, and therefore equipment sizes throughout the plant; this is reflected in the higher EPCC and TPC costs in Case 2a. On a $/kW basis, the TPC of the advanced "F" turbine plant is less because of increased power output.

Table 3-5. Case 2a: Capital and O&M Cost Comparison

	Case 0			Case 2a		
	Reference plant with 7FA			Reference plant with adv. "F"		
Capital Cost ($1,000)						
Plant Sections	**EPCC**	**TPC**	**TPC $/kW**	**EPCC**	**TPC**	**TPC $/kW**
1 Coal and Sorbent Handling	25,685	30,821	63	29,076	34,890	55
2 Coal and Sorbent Prep & Feed	39,472	48,980	101	45,169	56,050	88
3 Feedwater & Balance of Plant	28,606	35,077	72	30,636	37,513	59
4a Gasifier	184,371	236,212	485	210,196	269,284	423
4b Air Separation Unit	153,591	168,950	347	167,073	183,781	289

Table 3-5. (Continued)

	Case 0			Case 2a		
	Reference plant with 7FA			Reference plant with adv. "F"		
5a Gas Cleanup	93,441	112,389	231	107,769	129,625	203
5b CO$_2$ Removal & Compression	0	0	0	0	0	0
6 Gas Turbine	91,110	105,058	215	103,491	119,302	187
7 HRSG	44,560	49,511	102	50,936	56,565	89
8 Steam Cycle and Turbines	47,842	54,310	112	57,934	65,820	103
9 Cooling Water System	20,099	24.233	50	22,515	27,140	43
10 Waste Solids Handling System	34,981	38,752	80	39,568	43,829	69
11 Accessory Electric Plant	55,772	66,529	137	58,402	69,559	109
12 Instrumentation & Control	18,982	23,178	48	19,010	23,212	36
13 Site Preparation	13,956	18,143	37	14,247	18,522	29
14 Buildings and Structures	14,012	16,314	34	14,974	17,421	27
Total	866,482	1,028,457	2,113	970,995	1,152,513	1,809

O&M Cost ($1,000)						
Fixed Costs	**Total**		**% EPCC**	**Total**		**% EPCC**
Labor	19,542		2.26	22,548		2.32
Variable Operating Costs*	**Total**		**% EPCC**	**Total**		**% EPCC**
Maintenance Materials	18,368		2.12	21,339		2.20
Water	1,451		0.17	1,596		0.16
Chemicals	1,021		0.12	1,215		0.13
Waste Disposal	2,262		0.26	2,745		0.28
Total Variable Costs	23,102		2.67	26,896		2.77
Total O&M Cost	42,644		4.92	49,444		5.09
Fuel Cost*	55,690		6.43	68,008		7.00

Discounted Cash Flow Results						
Total Plant Cost ($/kW)			2,113			1,809
Levelized Cost of Electricity ($/kW-hr)			0.0927			0.0814

*Includes 75 % Capacity Factor

Comparing cost of electricity, the $0.0814/kW-hr of Case 2a is less than the $0.0927/kW-hr of Case 0 because of (1) larger gas turbine, which increases the plant output and therefore decreases the capital cost on a $/kW basis, and (2) increased plant efficiency due to the higher pressure ratio and firing temperature of the advanced "F" frame syngas turbine compared to the 7FA turbine.

3.3.2. Cumulative Impact of R&D

Composite Process Configuration (Case 2): Slurry Feed Gasifier, Cryogenic ASU, Cold Gas Cleanup, Advanced "F" Frame Syngas Turbine, 80 % Capacity Factor

The improvement from Case 1 to Case 2 is replacement of the 7FA gas turbine with the more advanced "F" frame gas turbine. Replacement of the turbine is already represented by Case 2a above, so the cumulative impact can be evaluated simply by increasing the capacity factor of Case 2a to 80 %. As a result, the net plant efficiency of Case 2 is still 37.9 % (the same as Case 2a). The only change is to increase the net plant operating hours, which decreases the levelized cost of electricity from $0.0814/kW-hr to $0.0780/kW-hr.

This case is nearly identical to Case 1 of NETL's Baseline Study; a validation was performed to demonstrate that results are consistent. The validation is presented in Appendix A.1.

3.4. Coal Feed Pump

The benefit of this technology is to decrease the energy required to evaporate slurry water in the gasifier, thereby increasing cold gas efficiency of the gasifier. Dry feed is accomplished with a coal pump, which is assumed to be capable of delivering dry feed to the elevated gasifier pressure. A fluffing gas is required for coal transport, and has an assumed flowrate of 0.156 lb. fluff gas per lb. coal feed. The power requirement for the coal pump is 500 kW per gasifier.

The coal pump is assumed capable of delivering as-received coal to the gasifier without the need for coal drying. The coal feed, with 12.5 % moisture, is considered to contain sufficient moisture that additional steam is not needed for gasification.

3.4.1. Impact of Coal Feed Pump in the Reference Plant (Case 3 a)

Case 3a Configuration: Coal Feed Pump, Cryogenic ASU, Cold Gas Cleanup, 7FA Syngas Turbine, 75 % Capacity Factor

The slurry feed gasifier is replaced with a coal feed pump that eliminates the need for slurry water, and the performance improvement is evaluated. The process configuration is identical to that in Figure 3-2, except that coal is delivered to the gasifier as dry feed rather than slurry. Dry feed has the advantage of less energy consumed in the gasifier to evaporate water from the slurry, resulting in a greater portion of the coal feed converted to CO (rather than CO_2) in the raw syngas and thereby increasing the cold gas efficiency of the gasifier.

The raw syngas composition in Case 3a has much less water because of the dry feed. Less coal is needed in this case, so the molar flowrate of raw syngas is also less. The concentration of CO in the Case 3a syngas is much greater – due to not having to oxidize carbon in the gasifier in order to evaporate slurry water.

With the 7FA reference plant gas turbine, neither case integrates air from the gas turbine to the ASU. Because of the decreased coal feedrate in Case 3a and corresponding decrease in gasifier oxygen required, all available N_2 from the ASU is used for fuel dilution and the fuel gas must also be humidified in order to generate sufficient flow through the gas turbine. Case 3a has, as a result, a higher mole fraction of H_2O in the fuel gas due to humidification.

Table 3-6 below summarizes the overall process performance for two process trains.

The reduced steam turbine power in Case 3a reflects decreased coal flowrate (as the result of not having to evaporate slurry water), leading to decreased heat recovery in the gasifier and syngas quench. Another factor adding to decreased steam turbine power in Case 3a is heat required to humidify the fuel stream.

The primary differences in auxiliary power consumption lay in the ASU main air compressor, the oxygen compressor, and Selexol auxiliaries. These all result from reduced coal feedrate to the gasifier in Case 3a, and also scale back all other auxiliary power accounts – reducing auxiliary power in Case 3a by 14 MW.

Overall, the net plant efficiency increases by 1.9 percentage points going from the Reference Case to the coal feed pump.

Cost Analysis (Case 3a)

Capital and O&M costs are compared with Case 0 results in Table 3-7. The coal feed pump Case 3a has a lower coal flowrate (due to not having to evaporate slurry water in the gasifier) and lower net power production; these are reflected in generally lower capital costs due to equipment scale factors. Some accounts, such as coal handling, coal prep, instrumentation, site preparation, and buildings, reduce slightly as a result. Other accounts, such as gasifier, ASU, and gas cleanup, have much more pronounced cost reductions as the result of less heat transfer in the radiant cooler, less oxygen demand in the gasifier, and less syngas to desulfurize. The net $74 million reduction in TPC translates to $65/kW (3 %) reduction in capital cost – due primarily to plant scale reductions made possible by the dry feed gasifier rather than the reduced cost of coal feed pump equipment vs. slurry feed equipment.

Table 3-6. Performance Impact of Coal Feed Pump in the Reference Plant

	Case 0	Case 3a
	Reference plant with slurry feed	**Reference plant with coal feed pump**
Gas Turbine Power (MWe)	384	384
Fuel Gas Expander (MWe)	6	7
Steam Turbine Power (MWe)	223	188
Total Power Produced (MWe)	614	579
Auxiliary Power Use (MWe)	-127	-113
Net Power (MWe)	487	466
As-Received Coal Feed (lb/hr)	402,581	365,931
Net Heat Rate (Btu/kW-hr)	9,649	9,157
Net Plant Efficiency (HHV)	35.4 %	37.3 %
Gasifier Cold Gas Efficiency	75.8 %	81.6 %

With little difference between total O&M cost between the Reference Case and the coal feed pump case, the primary operating cost reduction is in the fuel cost (which is made possible by the 1.9 percentage point improvement in process efficiency) of about $5 MM/yr.

The combined reductions in capital cost and fuel cost contribute to a reduction in levelized cost of electricity from $0.0927/kW-hr to $0.0894/kW-hr, or about a 3.6 percent reduction in COE due to the coal feed pump.

Table 3-7. Case 3a: Capital and O&M Cost Comparison

	Case 0			Case 3a		
	Reference plant with slurry feed			Reference plant with coal pump		
Capital Cost ($1,000)						
Plant Sections	**EPCC**	**TPC**	**TPC $/kW**	**EPCC**	**TPC**	**TP C $/kW**
1 Coal and Sorbent Handling	25,685	30,821	63	24,208	29,049	62
2 Coal and Sorbent Prep & Feed	39,472	48,980	101	38,364	47,373	102
3 Feedwater & Balance of Plant	28,606	35,077	72	25,291	30,961	66
4a Gasifier	184,371	236,212	485	167,010	214,134	459
4b Air Separation Unit	153,591	168,950	347	137,625	151,387	325
5a GasCleanup	93,441	112,389	231	85,697	103,057	221
5b CO2 Removal & Compression	0	0	0	0	0	0
6 Gas Turbine	91,110	105,058	215	91,228	105,195	226
7 HRSG	44,560	49,511	102	44,482	49,425	106
8 Steam Cycle and Turbines	47,842	54,3 10	112	42,318	48,016	103
9 Cooling Water System	20,099	24,2 33	50	18,405	2 2, 19 6	48
10 Waste Solids Handling System	34,981	38,7 52	80	31,179	34,54 3	74
11 Accessory Electric Plant	55,772	66,5 29	137	53,391	63,67 2	137
12 Instrumentation & Control	18,982	23,1 78	48	18,345	22,40 0	48
13 Si te Preparation	13,956	18, 143	37	13,701	17,81 2	38
14 B uildings and Structures	14,012	16,3 14	34	13,233	15,41 6	33
Total	86 6,482	1,02 8,457	2,11 3	804,477	954,636	2,048

Table 3-7. (Continued)

O&M Cost ($1,000)				
Fixed Costs	**Total**	**C**	**% EPC% Total**	**EPCC**
Labor	19,542	2.26	18,03 9	2.24
Variable Operating Costs*	**Total**	**% EPCC**	**Total**	**% E PCC**
Maintenance Materials	18,368	2.12	17,681	2.20
Water	1,451	0.17	1,092	0.14
Chemicals	1,021	0.12	967	0.12
Waste Disposal	2,262	0.26	1,886	0.23
Total Variable Costs	23,102	2.67	21,626	2.69
Total O&M Cost	42,644	4.92	39,665	4.93
Fuel Cost*	55,690	6.43	50,620	6.29
Discounted Cash Flow Results				
Total Plant Cost ($/kW)		2,113		2,048
Levelized Cost of Electricity ($/kW-hr)		0.0927		0.0894

*Includes 75 % capacity factor 3.4.2

3.4.2. Cumulative Impact of R&D

Composite Process Configuration (Case 3): Coal Feed Pump, Cryogenic ASU, Cold Gas Cleanup, Advanced "F" Frame Syngas Turbine, 80 % Capacity Factor

In the cumulative case, the combined performance from increased capacity factor to 80 %, replacement of the 7FA turbine with advanced "F" frame turbine, and coal feed pump is examined. A block flow diagram of this process is shown in Figure 3-3. The table below compares the incremental improvement due to the coal feed pump.

Table 3-8. Incremental Performance Improvement from the Coal Feed Pump

	Case 2	Case 3
	80% CF, adv. "F"	**80% CF, adv. "F", coal feed pump**
Gas Turbine Power (MWe)	464	464
Fuel Gas Expander (MWe)	8	8
Steam Turbine Power (MWe)	293	256
Total Power Produced (MWe)	765	728
Auxiliary Power Use (MWe)	-128	-114
Net Power (MWe)	637	614
As-Received Coal Feed (lb/hr)	491,633	449,270
Net Heat Rate (Btu/kW-hr)	9,004	8,542
Net Plant Efficiency (HHV)	37.9 %	40.0 %
Gasifier Cold Gas Efficiency	76.0 %	81.9 %

The oxygen:coal weight ratio in Case 2 was 0.94 lb O_2 / lb dry coal. In Case 3, the ratio is 0.86 lb O_2 / lb dry coal. The reduced O_2 requirement is reasonable for Case 3. Less oxidation will be required per pound of coal in Case 3 since there is no slurry water to evaporate. As a result, gasifier cold gas efficiency increases from 76.0 % in Case 2 to 81.9 % in Case 3.

Three reasons account for reduced steam turbine power generation in Case 3. First, Case 3 partially humidifies the fuel gas because there is not enough N_2 from the ASU to dilute the fuel gas to the 125 Btu/scf heating value specification. Second, there is less coal flowrate through the gasifier in Case 3 since there is no need to evaporate slurry water, and therefore less heat is available in the radiant c ooler for steam generation. Third, the flowrate of raw fuel gas in the quench system decreases in Case 3, reducing the amount of heat recovery. These factors account for 37 MW less steam turbine power generation in the case of the coal feed pump.

Auxiliary power consumption in Case 3 is 14 MW less than in Case 2 as the result of reduced oxygen demand and ASU air supply (reducing the power requirement of the main air compressor). Overall, the net power generated in Case 3 is 23 MW less than Case 2, but the c oal feed rate required to achieve the 232 MWe gas turbine rating is significantly lower – resulting in an improved net plant efficiency from 37.9 % to 40.0 %.

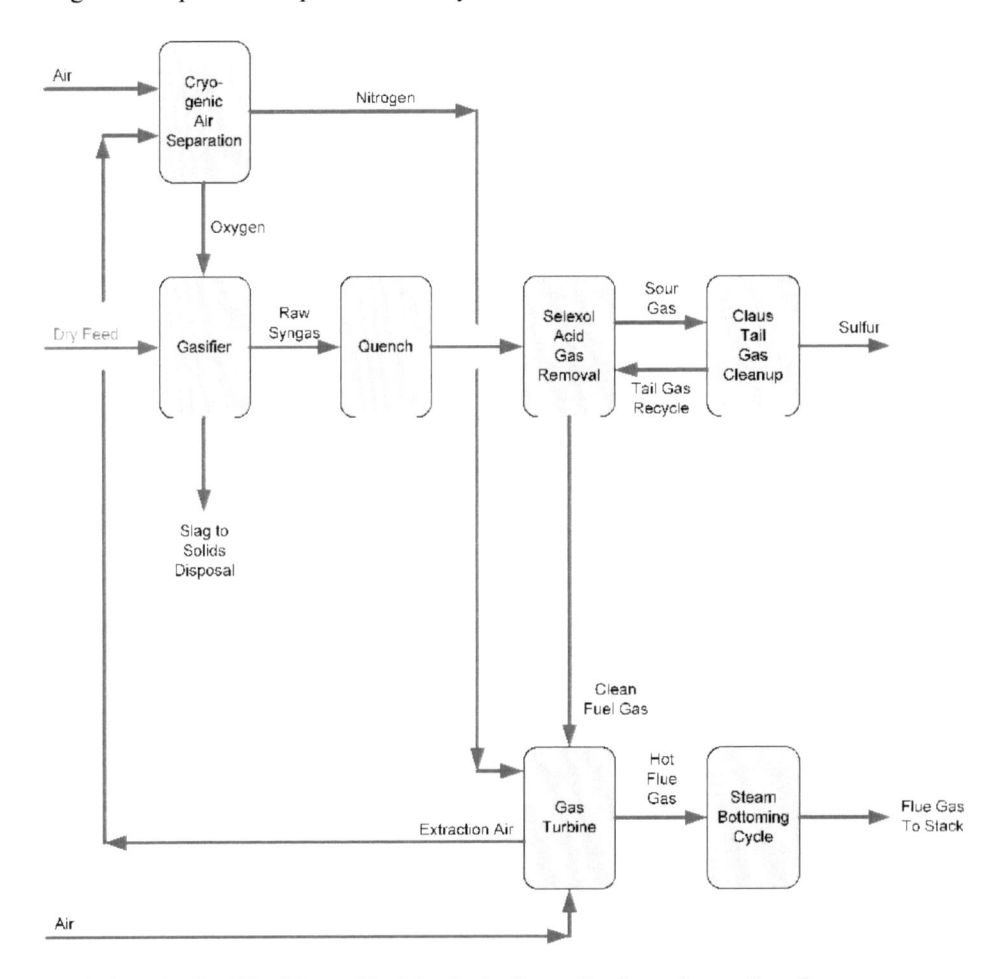

Figure 3-3. Case 3: Coal Feed Pump Has Nearly the Same Configuration as Case 2.

Cost Analysis (Case 3)

As shown in Table 3-9, cost accounts in Case 3 generally decrease due to smaller equipment size as the result of reduced coal flowrate. Such accounts include coal handling, coal prep, balance of plant (BOP), gas cleanup, steam cycle, cooling water, and solid waste handling.

Greater cost reductions in gasifier (due to reduced radiant cooler heat duty) and air separation unit (due to reduced oxygen demand) reflect the primary cost advantages of switching from slurry feed gasifier to dry feed.

The bottom-line cost reduction in total plant cost is about $80 million, or in other words a reduction of about $60/kW. This translates to a COE reduction from $0.0780/kW-hr to $0.0751/kW-hr, a savings of about 3.7 % in cost of electricity.

Table 3-9. Case 3: Capital and O&M Cost Comparison

	Case 2			Case 3		
	80% CF, adv. "F"			80% CF, adv. "F", coal pump		
Capital Cost ($1,000)						
Plant Sections	**EPCC**	**TPC**	**TPC $/kW**	**EPCC**	**TPC**	**TPC $/kW**
1 Coal and Sorbent Handling	29,076	34,890	55	27,494	32,993	54
2 Coal and Sorbent Prep & Feed	45,169	56,050	88	44,060	54,408	89
3 Feedwater & Balance of Plant	30,636	37,513	59	27,259	33,322	54
4a Gasifier	210,196	269,284	423	191,060	244,956	399
4b Air Separation Unit	167,073	183,781	289	148,372	163,209	266
5a Gas Cleanup	107,769	129,625	203	99,131	119,216	194
5b CO_2 Removal & Compression	0	0	0	0	0	0
6 Gas Turbine	103,491	119,302	187	103,552	119,373	195
7 HRSG	50,936	56,565	89	51,276	56,942	93
8 Steam Cycle and Turbines	57,934	65,820	103	52,627	59,769	97
9 Cooling Water System	22,515	27,140	43	20,994	25,313	41
10 Waste Solids Handling System	39,568	43,829	69	35,385	39,199	64
11 Accessory Electric Plant	58,402	69,559	109	56,192	66,908	109
12 Instrumentation & Control	19,010	23,212	36	18,430	22,503	37
13 Site Preparation	14,247	18,522	29	13,995	18,193	30
14 Buildings and Structures	14,974	17,421	27	14,244	16,579	27
Total	970,995	1,152,513	1,809	904,070	1,072,883	1,749

Table 3-9. (Continued)

O&M Cost ($1,000)				
Fixed Costs	**Total**	**% EPCC**	**Total**	**% EPCC**
Labor	22,548	2.32	21,045	2.33
Variable Operating Costs	**Total**	**% EPCC**	**Total**	**% EPCC**
Maintenance Materials	22,762	2.34	21,993	2.43
Water	1,703	0.18	1,283	0.14
Chemicals	1,305	0.13	1,235	0.14
Waste Disposal	2,920	0.30	2,451	0.27
Total Variable Costs	28,694	2.96	26,961	2.98
Total O&M Cost	51,237	5.28	48,006	5.31
Fuel Cost	72,542	7.47	66,291	7.33
Discounted Cash Flow Results				
Total Plant Cost ($/kW)		1,809		1,749
Levelized Cost of Electricity ($/kW-hr)		0.0780		0.0751

3.5. Increased Capacity Fact or to 85 Percent

In Case 4, the process configuration and process performance remains the same as Case 3, but the capacity factor increases from 80 percent to 85 percent. The increased power production resulting from more time on-line reflects anticipated improvements in process reliability, availability, and maintainability (RAM) due to DOE-sponsored R&D in areas such as vessel refractories and improved sensors (with no additional capital or fixed O&M cost).

The differences between Case 3 and Case 4 lie in variable O&M costs, fuel cost, and plant revenues as the result of longer hours of operation. Variable O&M costs increase by about $1.7 MM/year, and fuel costs increase by about $4.1 MM/year. The increased plant revenue from additional power production results in decreased cost of electricity from $0.0751/kW-hr in Case 3 to $0.0722/kW-hr in Case 4 – a savings of about 3.9 % in cost of electricity resulting from increased capacity factor.

3.6. Transport Desulfurizer and Direct Sulfur Reduction Process

In Case 5, the primary process improvement is that the Selexol (cold gas cleanup) acid gas removal and Claus tail gas treatment processes are replaced with warm gas Transport Desulfurizer (TDS) and Direct Sulfur Reduction (DSRP) processes. Exiting the gasifier, raw syngas is cooled to approximately 950 °F in preparation for hydrogen chloride removal in a packed bed of Na_2CO_3 (trona). Hydrogen chloride is removed according to the reaction:

$$2 \, HCl + Na_2CO_3 \, (s) = 2 \, NaCl \, (s) + CO_2 + H_2O$$

The syngas is cooled in preparation for contact with zinc oxide sorbent, which reacts with H_2S and COS to remove them from the syngas. The desulfurization reactions are:

$$H_2S + ZnO\ (s) = ZnS\ (s) + H_2O$$

$$COS + ZnO\ (s) = ZnS\ (s) + CO_2$$

Desulfurized syngas is cooled to about 150 °F in preparation for cold gas ammonia and mercury removal. Ammonia is removed by scrubbing with water. An activated carbon filter bed is used for mercury removal. Clean fuel gas is reheated before expansion through the fuel gas expander.

To regenerate the ZnO sorbent for the TDS, ZnS transfers to the TDS regenerator where it contacts with air and is oxidized at 1,100 °F according to the reaction:

$$ZnS\ (s) + 3/2\ O_2 = ZnO\ (s) + SO_2$$

The SO_2 (sour gas) that is generated flows to the DSRP for sulfur recovery, and the regenerated sorbent is returned to the transport desulfurizer. A small portion of clean fuel gas exiting the mercury removal section is used as reducing gas in the DSRP where sour gas is reduced, forming elemental sulfur product:

$$SO_2 + 2\ H_2 = 2\ H_2O + S$$

$$SO_2 + 2\ CO = 2\ CO_2 + S$$

DSRP tail gas, containing H_2O and CO_2, is compressed and recycled to the transport desulfurizer. The transport desulfurizer has the advantage of eliminating solvent regeneration (and therefore steam heat duty) in the Selexol reboiler. Instead, acid gas, deposited on a solid zinc oxide sorbent, is oxidized during sorbent regeneration.

While elimination of the Selexol reboiler reduces steam consumption, oxidation during zinc oxide sorbent regeneration actually produces heat – both effects contributing to increased steam power generation and therefore increased energy efficiency.

3.6.1. Impact of Partial Warm Gas Cleanup in the Reference Plant (Case 5a)

Case 5a Configuration: Slurry Feed Gasifier, Cryogenic ASU, Transport Desulfurizer and DSRP, 7FA Syngas Turbine, 75 % Capacity Factor

Table 3-10 below compares overall performance when the partial warm gas cleanup process is implemented in the reference plant.

The increased steam turbine power by 42 MW in Case 5a reflects elimination of the Selexol reboiler and the sour water stripper reboiler, and recovery of high quality heat from the clean syngas exiting the transport desulfurizer (as opposed to reheat of clean fuel gas prior to thefuel gas expander).

The regeneration air compressor auxiliary power is a trade-off for elimination of the Selexol Unit Auxiliaries, which slightly increases auxiliary power use in warm gas cleanup by 2 MW. Auxiliary power accounts associated with the steam cycle and net plant power output are slightly greater in Case 5a due to the increased steam turbine power output in that case. Overall, auxiliary power increases in the transport desulfurizer case by about 2 MW.

Table 3-10. Performance Impact of Partial Warm Gas Cleanup in the Reference Plant

	Case 0	Case 5a
	Reference plant with cold gas cleanup	Reference plant with partial warm gas cleanup
Gas Turbine Power (MWe)	384	384
Fuel Gas Expander (MWe)	6	7
Steam Turbine Power (MWe)	223	265
Total Power Produced (MWe)	614	656
Auxiliary Power Use (MWe)	-127	-129
Net Power (MWe)	487	527
As-Received Coal Feed (lb/hr)	402,581	412,206
Net Heat Rate (Btu/kW-hr)	9,649	9,123
Net Plant Efficiency (HHV)	35.4 %	37.4 %

Table 3-11. Case 5a: Capital and O&M Cost Comparison

	Case 0			Case 5a		
	Reference plant with cold gas cleanup			Reference plant with partial warm gas cleanup		
Capital Cost ($1,000)						
Plant Sections	EPCC	TPC	TPC $/kW	EPCC	TPC	TPC $/ kW
1 Coal and Sorbent Handling	25,685	30,821	63	26,064	31,277	59
2 Coal and Sorbent Prep & Feed	39,472	48,980	101	40,105	49,766	94
3 Feedwater & Balance of Plant	28,606	35,077	72	28,834	35,351	67
4a Gasifier	184,371	236,212	485	185,977	237,034	450
4b Air Separation Unit	153,591	168,950	347	153,726	169,098	321
5a Gas Cleanup	93,441	112,389	231	64,675	77,975	148
5b CO_2 Removal & Compression	0	0	0	0	0	0
6 Gas Turbine	91,110	105,058	215	91,438	105,437	200
7 HRSG	44,560	49,511	102	44,592	49,546	94
8 Steam Cycle and Turbines	47,842	54,310	112	54,049	61,385	116
9 Cooling Water System	20,099	24,233	50	21,801	26,280	50
10 Waste Solids Handling System	34,981	38,752	80	35,497	39,323	75
11 Accessory Electric Plant	55,772	66,529	137	56,820	67,754	129

Table 3-11. (Continued)

	Case 0			Case 5a		
	Reference plant with cold gas cleanup			Reference plant with partial warm gas cleanup		
12 Instrumentation & Control	18,982	23,178	48	19,089	23,309	44
13 Site Preparation	13,956	18,143	37	14,009	18,211	35
14 Buildings and Structures	14,012	16,314	34	14,617	17,010	32
Total	866,482	1,028,457	2,113	851,292	1,008,754	1,913

O&M Cost ($1,000)

Fixed Costs	Total		% EPCC	Total		% EPCC
Labor	19,542		2.26	19,542		2.30

Variable Operating Costs*	Total		% EPCC	Total		% EPCC
Maintenance Materials	18,368		2.12	19,215		2.26
Water	1,451		0.17	1,430		0.17
Chemicals	1,021		0.12	3,790		0.45
Waste Disposal	2,262		0.26	2,314		0.27
Total Variable Costs	23,102		2.67	26,749		3.14
Total O&M Cost	42,644		4.92	46,291		5.44
Fuel Cost*	55,690		6.43	57,021		6.70

Discounted Cash Flow Results

Total Plant Cost ($/kW)	2,113	1,913
Levelized Cost of Electricity ($/kW-hr)	0.0927	0.0862

*Includes 75 % capacity factor

Because of the large increase in steam turbine power generation, net plant efficiency increases by 2.0 percentage points by replacing the Reference Case cold gas cleanup with partial warm gas cleanup consisting of transport desulfurizer with chloride guard bed and DSRP.

Cost Analysis (Case 5a)

Capital and O&M costs are compared with Reference Case results in Table 3-11. The transport desulfurizer case has significantly greater net power production (527 MW vs. 487 MW) resulting from slightly greater coal feedrate and 2.0 percentage point increase in process efficiency; these translate to higher TPC in most capital cost accounts (but TPC decreases on a $/kW basis). Notable exceptions are the gas cleanup section (which replaces cold gas cleanup process equipment with less expensive warm gas cleanup equipment – resulting in a cost savings of $34 million) and the steam cycle (which has an increased power production of 42 MW, resulting in a cost increase of $7 million). Overall, there is a net decrease in TPC of $20 million; this translates to a decrease of $200/kW, or a 9.5 % decrease in capital cost on a $/kW basis.

The cost of chemicals increases significantly from the Reference Case due to the cost of ZnO sorbent and trona. A slight attrition of sorbent (10-20 lb per million lb of sorbent recirculation) is assumed; fresh sorbent replacement cost is assumed to be $14,000/ton. The cost of trona is calculated as twice the stoichiometric quantity required to convert chloride, at a cost of $1 00/ton.

As a result of increased chemicals cost, total variable costs in Case 5a increase by about $3.6 MM/year. Fuel cost increases by about $1.3 MM/year due to the slightly increased coal feed rate. The decreased TPC and increased net power more than compensate for the increased operating costs, however, resulting in a levelized COE reduction from $0.0927/kW-hr in the Reference Case to $0.0862/kW-hr in Case 5a – a reduction by 7.0 % in cost of electricity resulting from the transport desulfurizer with chloride guard bed and DSRP.

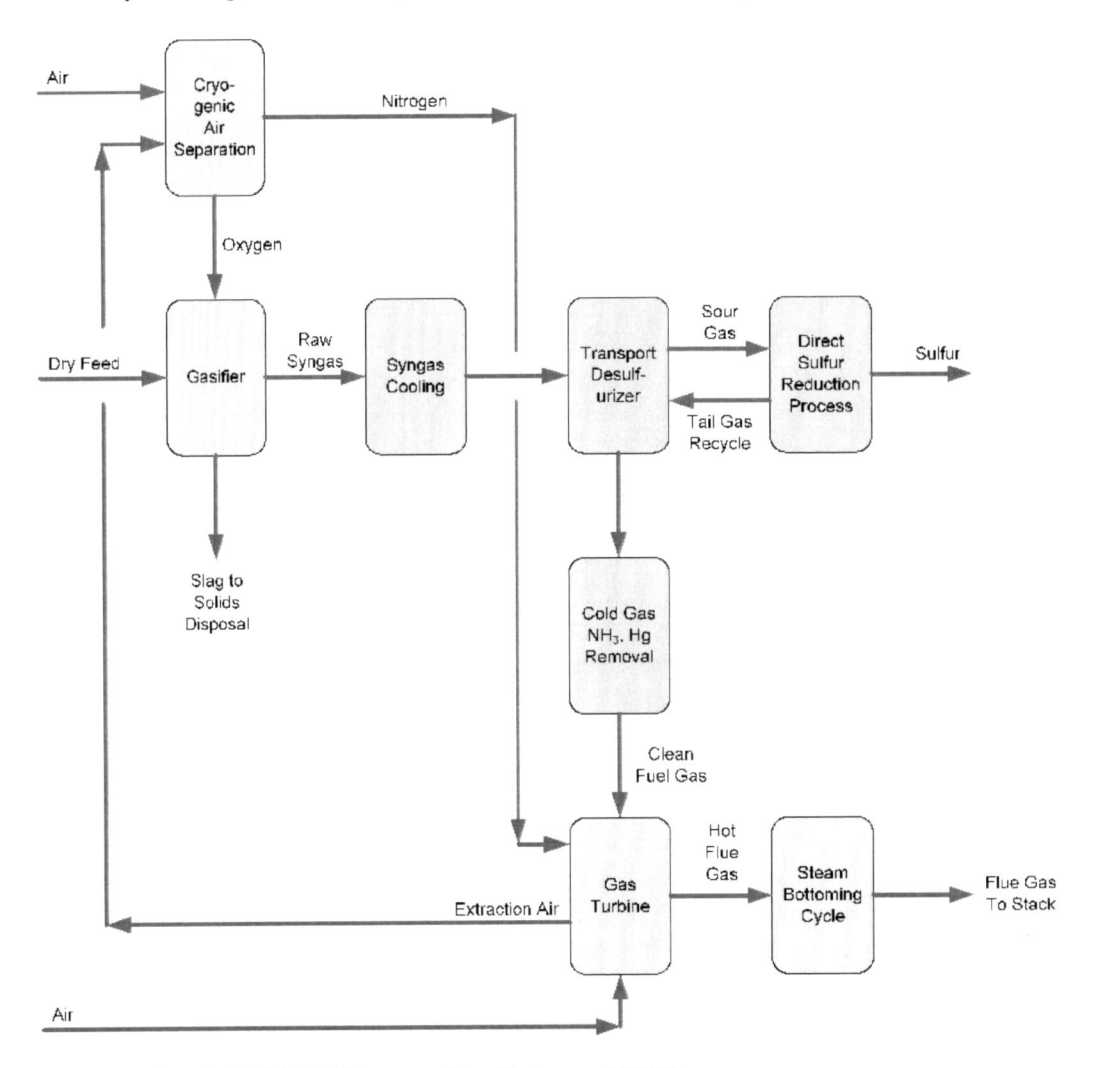

Figure 3-4. Case 5: IGCC With Transport Desulfurizer and DSRP.

3.6.2. Cumulative Impact of R&D

Composite Process Configuration (Case 5): Coal Feed Pump, Cryogenic ASU, Transport Desulfurizer and DSRP, Advanced "F" Frame Syngas Turbine, 85 % Capacity Factor

The block flow diagram in Figure 3-4 shows the implementation of partial warm gas cleanup in the composite process with other advanced technology.

Overall process performance and the improvement in net plant efficiency due to warm gas desulfurization in the composite process are shown in Table 3-12.

Case 5 generates significantly more steam turbine power (49 MW) as a result of eliminating the Selexol and sour water stripper reboilers, and less humidifying steam is needed in Case 5. Auxiliary power consumption increases slightly because the increase in regeneration air compressor power outweighs the savings in Selexol unit auxiliaries. The larger bottoming cycle in Case 5 also contributes to greater auxiliary power use.

Overall, the net power production is greater in Case 5 due to greater power recovered by the steam cycle. Although the coal feed rate required to achieve the 232 MWe gas turbine rating is somewhat greater, the additional steam power generation results in improved net plant efficiency from 40.0 % to 42.1 %.

Cost Analysis (Case 5)

Comparing TPC between Cases 4 and 5 in Table 3-13, the primary cost reduction occurs in the Gas Cleanup account. The reduction in that account is estimated at approximately $45 MM, or a reduction of $81/kW resulting from less expensive equipment.

The increased steam power production in Case 5 (by 49 MW) is responsible for increased steam cycle cost (by $8 MM). However, the increased net power production in Case 5 (by 45 MW) results in a consistent decrease in Total Plant Cost on a $/kW basis in all other accounts. The bottom-line cost reduction in total plant cost is almost $29 MM, resulting in a decrease of $164/kW.

Table 3-12. Incremental Performance Improvement from Partial Warm Gas Cleanup

	Case 4	Case 5
	Adv. "F", coal pump, 85 % CF	Adv. "F", coal pump, 85 % CF, partial WGCU
Gas Turbine Power (MWe)	464	464
Fuel Gas Expander (MWe)	8	8
Steam Turbine Power (MWe)	256	305
Total Power Produced (MWe)	728	777
Auxiliary Power Use (MWe)	-114	-119
Net Power (MWe)	614	659
As-Received Coal Feed (lb/hr)	449,270	457,603
Net Heat Rate (Btu/kW-hr)	8,542	8,105
Net Plant Efficiency (HHV)	40.0 %	42.1 %

Table 3-13. Case 5: Capital and O&M Cost Comparison

	Case 4 Adv. "F", coal pump, 85% CF			Case 5 Adv. "F", coal pump, 85% CF, partial WGCU		
Capital Cost ($1,000)						
Plant Sections	**EPCC**	**TPC**	**TPC $/kW**	**EPCC**	**TPC**	**TPC $/kW**
1 Coal and Sorbent Handling	27,494	32,993	54	27,810	33,372	51
2 Coal and Sorbent Prep & Feed	44,060	54,408	89	44,609	55,390	84
3 Feedwater & Balance of Plant	27,259	33,322	54	27,447	33,548	52
4a Gasifier	191,060	244,956	399	193,291	247,825	376
4b Air Separation Unit	148,372	163,209	266	146,987	161,685	245
5a Gas Cleanup	99,131	119,216	194	61,645	74,238	113
5b CO$_2$ Removal & Compression	0	0	0	0	0	0
6 Gas Turbine	103,552	119,373	195	103,621	119,452	181
7 HRSG	51,276	56,942	93	51,246	56,910	86
8 Steam Cycle and Turbines	52,627	59,769	97	59,672	67,803	103
9 Cooling Water System	20,994	25,313	41	22,844	27,537	42
10 Waste Solids Handling System	35,385	39,199	64	35,787	39,644	60
11 Accessory Electric Plant	56,192	66,908	109	57,541	68,497	104
12 Instrumentation & Control	18,430	22,503	37	18,626	22,743	35
13 SitePreparation	13,995	18,193	30	14,063	18,282	28
14 Buildings and Structures	14,244	16,579	27	14,923	17,361	26
Total	904,070	1,072,883	1,749	880,114	1,044,287	1,585
O&M Cost ($1,000)						
Fixed Costs	**Total**		**% EPCC**	**Total**		**% EPCC**
Labor	21,045		2.33	21,045		2.39
Variable Operating Costs	**Total**		**% EPCC**	**Total**		**% EPCC**
Maintenance Materials	23,368		2.59	24,471		2.78
Water	1,363		0.15	1,333		0.15
Chemicals	1,312		0.15	4,794		0.55
Waste Disposal	2,604		0.29	2,650		0.30
Total Variable Costs	28,646		3.17	33,249		3.78
Total O&M Cost	49,691		5.50	54,294		6.17
Fuel Cost	70,435		7.79	71,741		8.15
Discounted Cash Flow Results						
Total Plant Cost ($/kW)			1,749			1,585
Levelized Cost of Electricity ($/kW-hr)			0.0722			0.0677

The use of ZnO sorbent and trona increases the cost of chemicals significantly, but is a relatively small contribution to overall process economics. The transport desulfurizer with DSRP reduces COE from $0.0722/kW-hr to $0.0677/kW-hr – a savings of about 6.2 % in cost of electricity.

Table 3-14. Performance Impact of Full Warm Gas Cleanup in the Reference Plant

	Case 0	Case 5a	Case 6a
	Reference plant with CGCU	Reference plant with partial WG CU	Reference plant with full WGCU
Gas Turbine Power (MWe)	384	384	384
Fuel Gas Expander (MWe)	6	7	9
Steam Turbine Power (MWe)	223	265	270
Total Power Produced (MWe)	614	657	663
Auxiliary Power Use (MWe)	-127	-129	-122
Net Power (MWe)	487	527	541
As-Received Coal Feed (lb/hr)	402,581	412,206	414,455
Net Heat Rate (Btu/kW-hr)	9,649	9,123	8,933
Net Plant Efficiency (HHV)	35.4 %	37.4 %	38.2 %

3.7. Full Warm Gas Cleanup

For full warm gas cleanup, novel treatment systems for ammonia and mercury removal are added to the chloride guard bed, transport desulfurizer, and DSRP. Warm gas ammonia removal eliminates direct contact cooling with water to remove NH3 downstream of the transport desulfurizer – effectively retaining moisture in the fuel gas because temperature remains above the dew point. For a fuel gas containing significant moisture, this has the benefit of maintaining flow through the fuel gas expander, and requires less dilution nitrogen in the topping combustor (which decreases auxiliary power consumption).

3.7.1. Impact of Full Warm Gas Cleanup in the Reference Plant (Case 6a)

Case 6a Configuration: Slurry Feed Gasifier, Cryogenic ASU, Warm Gas Cleanup, 7FA Syngas Turbine, 75 % Capacity Factor

Results are compared against the cold gas cleanup technology of Case 0 and cold gas ammonia and mercury removal process steps of Case 5a in order to evaluate the incremental contribution of the novel warm gas ammonia and mercury removal technologies.

Because the clean fuel in Case 6a retains moisture, flowrate is greater and the power generated by the fuel gas expander increases by 2 MW over Case 5a. The increased 5 MW of steam turbine power in Case 6a reflects (1) improved thermal efficiency within the novel ammonia and mercury removal section as the result of not cooling to as low a temperature and reheating the syngas, and (2) the greater turbine exhaust temperature (and therefore HRSG inlet temperature) in Case 6a resulting from higher moisture content in that flue gas.

The primary difference in auxiliary power consumption is due to the nitrogen compressor. Case 6a retains all moisture in the syngas, and therefore a smaller amount of nitrogen from the ASU is injected to reach the design power rating of the gas turbine. This results in a net 7 MW decrease in auxiliary power consumption in Case 6a.

Table 3-15. Case 6a: Capital and O&M Cost Comparison

	Case 5a			Case 6a		
	Reference plant with partial WGCU			Reference plant with full WGCU		
Capital Cost ($1,000)						
Plant Sections	**EPCC**	**TPC**	**TPC $/kW**	**EPCC**	**TPC**	**TPC $/kW**
1 Coal and Sorbent Handling	26,064	31,277	59	26,150	31,380	58
2 Coal and Sorbent Prep & Feed	40,105	49,766	94	40,253	49,948	92
3 Feedwater & Balance of Plant	28,834	35,351	67	28,887	35,415	65
4a Gasifier	1 85,977	237,034	450	186,644	237,883	440
4b Air Separation Unit	153,726	169,098	321	154,255	169,681	314
5a Gas Cleanup	64,675	77,975	148	70,804	85,484	158
5b CO_2 Removal & Compression	0	0	0	0	0	0
6 Gas Turbine	91,438	105,437	200	91,852	105,917	195
7 HRSG	44,592	49,546	94	44,943	49,931	92
8 Steam Cycle and Turbines	54,049	61,385	116	54,744	62,178	115
9 Cooling Water System	21,801	26,280	50	21,985	26,501	49
10 Waste Solids Handling System	35,497	39,323	75	35,615	39,453	73
11 Accessory Electric Plant	56,820	67,754	129	56,066	66,828	123
12 Instrumentation & Control	19,089	23,309	44	18,774	22,924	42
13 Site Preparation	14,009	18,211	35	13,953	18,139	34
14 Buildings and Structures	14,617	17,010	32	14,638	17,033	31
Total	851,292	1,008,754	1,913	859,564	1,018,696	1,882
O&M Cost ($1,000)						
Fixed Costs	**Total**		**% EPCC**	**Total**		**% EPCC**
Labor	19,542		2.30	19,542		2.27
Variable Operating Costs*	**Total**		**% EPCC**	**Total**		**% EPCC**
Maintenance Materials	19,215		2.26	19,346		2.25
Water	1,430		0.17	1,441		0.17
Chemicals	3,790		0.45	3,812		0.44
Waste Disposal	2,314		0.27	2,327		0.27
Total Variable Costs	26,749		3.14	26,926		3.13
Total O&M Cost	46,291		5.44	46,468		5.40
Fuel Cost*	57,021		6.70	57,332		6.67
Discounted Cash Flow Results						
Total Plant Cost ($/kW)			1,913			1,882
Levelized Cost of Electricity ($/kW-hr)			0.0862			0.0846

*Includes 75 % capacity factor

Overall, Case 6a produces 14 MW more power with only a slight increase in coal feed rate, increasing net plant efficiency from 37.4 % to 38.2 %. Compared to Case 0, the total net plant efficiency due to full warm gas cleanup improves from 35.4 % to 38.2 % -- an increase by 2.8 percentage points.

Cost Analysis (Case 6a)

Capital and O&M costs are compared with Case 5a results in Table 3-15. Because of (1) greater net power production (541 MW vs. 527 MW), and (2) 0.8 percentage point increase in process efficiency, most capital cost accounts increase slightly. The only notable exception is the gas cleanup section (which replaces the wet ammonia scrubber with higher temperature ammonia and mercury removal systems) – resulting in increased plant section cost by $7 million and total plant cost by $10 million. However, because of the increased net power production, total plant cost on a $/kW basis reduces from $1,913/kW to $1,882/kW – a decrease by $31/kW or 1.6 % going from partial warm gas cleanup to full warm gas cleanup.

O&M and fuel costs are relatively unaffected by the change in process configuration, so the difference in levelized COE is due almost entirely to the change in total plant cost. The change, from $0.0862/kW-hr to $0.0846/kW-hr, represents a $0.0016/kW-hr or 1.9 % reduction in COE due to the novel ammonia and mercury removal processes.

The overall reduction from the Reference Case 0 with cold gas cleanup ($2,1 13/kW capital and $0.0927/kW-hr COE) to full warm gas cleanup Case 6a ($ 1,882/kW capital and $0.0846/kW-hr COE) represents a 10.9 % decrease in capital cost and 8.7 % decrease in cost of electricity based on 75 % capacity factor.

Nexant [6] reported a 3.6 percentage point process efficiency improvement from replacing cold gas cleanup with warm gas cleanup in an IGCC process with a slurry feed gasifier; this was somewhat greater than the 2.8 percentage point improvement between Noblis' Cases 0 and 6a. In a separate analysis [7], Noblis determined that the difference in process efficiency was due to a series of design features. The two most significant of these were:

- The quantity of reducing gas sent to the SCOT process for tail gas treatment and the ultimate disposition of the tail gas (whether discarded or recycled to Selexol).
- Utilization of low-quality heat from the low-temperature syngas cooling section – i.e. the assumed temperature cut-off at which heat was considered to be unrecoverable for purposes of steam generation.

When changes were made to the Noblis configuration, Nexant's results could be reproduced. This highlights the importance of defining the baseline when quoting technology improvements. A validation of costs also showed that the Noblis and Nexant estimates for cost reduction were of similar magnitude. Appendix A.2 describes the results validation.

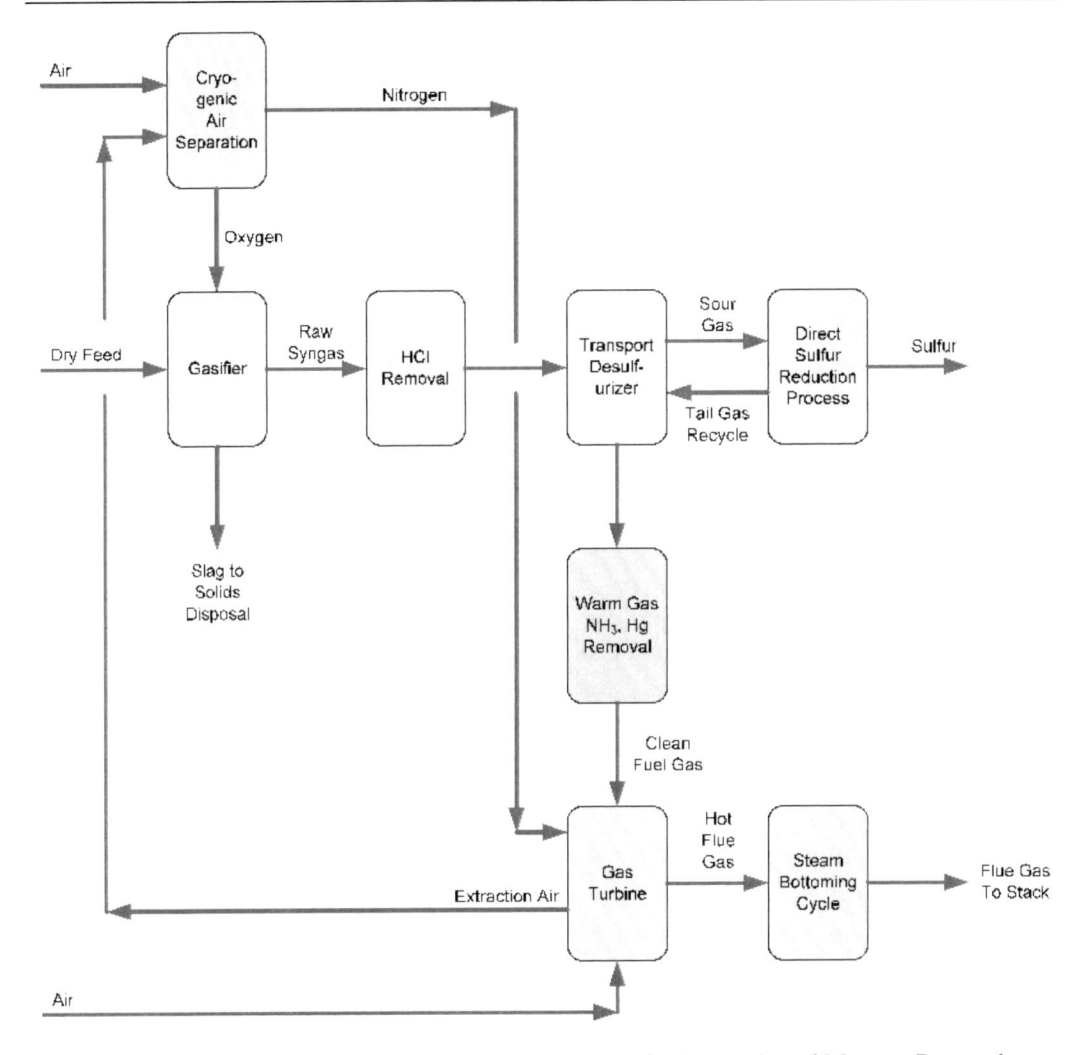

Figure 3-5. Case 6: IGCC With Novel Warm Gas Treatment for Ammonia and Mercury Removal.

3.7.2. Cumulative Impact of R&D

Composite Process Configuration (Case 6): Coal Feed Pump, Cryogenic ASU, Warm Gas Cleanup, Advanced "F" Frame Syngas Turbine, 85 % Capacity Factor

Figure 3-5 shows a block flow diagram of this process configuration. A key factor in the comparison between Cases 5 and 6 is the coal feed pump – as opposed to the slurry feed gasifier in the comparisons above between Cases 5a and 6a. Moisture from slurry water in Case 5a condenses during the quench for ammonia removal, but stays in the clean fuel gas in Case 6a thus increasing fuel flow and reducing the needed amount of nitrogen dilution.

Cases 5 and 6, on the other hand, have no appreciable moisture content because they have coal feed pumps, and therefore it doesn't make as much difference whether ammonia and mercury are removed at warm or cold temperature. Comparisons in Table 3-16 illustrate this.

Consistent with the comparison between Cases 5a and 6a, the slightly higher steam turbine power in Case 6 represents the effects of (1) not having to cool and reheat the syngas for NH_3 and mercury removal, and (2) having to humidify slightly less than in Case 5 to meet fuel specifications. However the auxiliary power consumption – and specifically the N_2 compressor work – is the same between Cases 5 and 6 when there was a difference of 7 MW between Cases 5a and 6a. This reduces the incremental process efficiency improvement to only 0.3 percentage points between Cases 5 and 6, as opposed to the 0.8 percentage point improvement between Cases 5a and 6a. The novel ammonia and mercury removal sections have a greater impact on the slurry feed gasifier process.

Table 3-16. Incremental Performance Improvement from Full Warm Gas Cleanup

	Case 5	Case 6
	Adv. "F", coal pump, 85% CF, partial WGCU	Adv. "F", coal pump, 85% CF, full WGCU
Gas Turbine Power (MWe)	464	4 64
Fuel Gas Expander (MWe)	8	9
Steam Turbine Power (MWe)	305	3 10
Total Power Produced (MWe)	777	7 83
Auxiliary Power Use (MWe)	-119	-119
Net Power (MWe)	659	6 64
As-Received Coal Feed (lb/hr)	457,603	457,710
Net Heat Rate (Btu/kW-hr)	8,103	8,042
Net Plant Efficiency (HHV)	42.1 %	42.4 %

Table 3-17. Case 6: Capital and O&M Cost Comparison

	Case 5			Case 6		
	Advanced "F", coal pump, 85% CF, partial WGCU			Advanced "F", coal pump, 85% CF, full WGCU		
Capital Cost ($1,000)						
Plant Sections	EPCC	TPC	TPC $/kW	EPCC	TPC	TPC $/kW
1 Coal and Sorbent Handling	27,810	33,372	51	27,811	33,373	50
2 Coal and Sorbent Prep & Feed	44,609	55,390	84	44,617	55,096	83
3 Feedwater & Balance of Plant	27,447	33,548	52	27,451	33,552	51
4a Gasifier	193,291	247,825	376	193,316	247,855	373
4b Air Separation Unit	146,987	161,685	245	146,987	161,686	244
5a Gas Cleanup	61,645	74,238	113	68,801	82,937	125
5b CO_2 Removal & Compression	0	0	0	0	0	0
6 Gas Turbine	103,621	119,452	181	103,747	119,599	180
7 HRSG	51,246	56,910	86	51,262	56,927	86
8 Steam Cycle and Turbines	59,672	67,803	103	60,341	68,566	103

Table 3-17. (Continued)

	Case 5 Advanced "F", coal pump, 85% CF, partial WGCU			Case 6 Advanced "F", coal pump, 85% CF, full WGCU		
9 Cooling Water System	22,844	27,537	42	23,016	27,744	42
10 Waste Solids Handling System	35,787	39,644	60	35,793	39,651	60
11 Accessory Electric Plant	57,541	68,497	104	57,657	68,632	103
12 Instrumentation & Control	18,626	22,743	35	18,636	22,755	34
13 Site Preparation	14,063	18,282	28	14,063	18,282	28
14 Buildings and Structures	14,923	17,361	26	14,988	17,435	26
Total	880,114	1,044,287	1,585	888,486	1,054,090	1,588

O&M Cost ($1,000)

Fixed Costs	Total		% EPCC	Total		% EPCC
Labor	21,045		2.38	21,045		2.37
Variable Operating Costs	**Total**		**% EPCC**	**Total**		**% EPCC**
Maintenance Materials	24,471		2.78	24,594		2.77
Water	1,333		0.15	1,343		0.15
Chemicals	4,794		0.55	4,808		0.54
Waste Disposal	2,650		0.30	2,651		0.30
Total Variable Costs	33,249		3.78	33,397		3.76
Total O&M Cost	54,294		6.16	54,442		6.13
Fuel Cost	71,741		8.15	71,758		8.08

Discounted Cash Flow Results

Total Plant Cost ($/kW)	1,585	1,588
Levelized Cost of Electricity ($/kW-hr)	0.0677	0.0675

Cost Analysis (Case 6)

Comparing total plant costs between Cases 5 and 6 in Table 3-17, the primary cost difference occurs in the gas cleanup account, which increases by about $9 MM. This is caused by (1) a slight increase in cost of mercury removal, and (2) a slight increase in cost of the novel ammonia removal system over the ammonia scrubber. The increase in the gas cleanup account is equivalent to about $ 12/kW; there is no significant difference in bottom-line total plant cost between the two cases. Likewise, there is no significant difference in operating costs; net power in Case 6 increases by less than 1 %, and the coal feed rates are very nearly equal. There is no significant COE improvement from the novel ammonia and mercury removal processes because neither the total plant cost nor the O&M cost changes.

It is interesting to compare this result with the difference between Cases 5a and 6a. Recall that in those cases, TPC decreases by $31/kW and COE decreases by $0.0016/kW-hr. The size of the gas turbine, and therefore plant power production, is greater in Cases 5 and 6; this increases the TPC and, when divided by the larger net power values, brings the $/kW values much closer together. To elaborate further, the incremental gas cleanup cost due to ammonia and mercury removal in Cases 5a and 6a is $7.5 MM; in Cases 5 and 6, it is $7.4 MM (it is reasonable that the incremental cost in the coal feed pump cases is less because of smaller fuel gas volume in the warm gas cleanup scenario). The incremental cost of gas cleanup carries through to the bottom-line TPC – resulting in $9.9 MM incremental cost from Case 5a to 6a, and an incremental cost of $8.5 MM from Case 5 to 6. The capital cost impacts by the novel ammonia and mercury removal sections are similar, but when divided by net power output there is a more noticeable difference in the smaller capacity plant – resulting in a $30/kW reduction in the smaller plant as opposed to negligible increase in the larger plant.

In conclusion, most of the performance and cost improvements of warm gas cleanup result from the transport desulfurizer, DSRP, and chloride guard bed. The incremental benefits from novel ammonia and mercury removal in a dry feed gasifier scenario appear to be minor.

3.8. Advanced Syngas Turbine – 2010-AST

DOE sponsors R&D to develop, by 2010, advanced syngas turbine technology with improved performance efficiency. Performance improvements are expected from higher pressure ratio and turbine inlet temperature, which will improve efficiency of the turbine.

3.8.1. Impact of 2010-AST Syngas Turbine in the Reference Plant (Case 7a)

Case 7a Configuration: Slurry Feed Gasifier, Cryogenic ASU, Cold Gas Cleanup, 2010-AST Syngas Turbine, 75 % Capacity Factor

The process block flow diagram of the reference IGCC process with a 2010-AST syngas turbine is identical to Figure 3-2 from Case 2a. Like the advanced "F" frame syngas turbine, the 2010-AST produces more power, has a higher pressure ratio, and higher firing temperature than the 7FA syngas turbine. In order to protect business-sensitive information, 2010-AST turbine performance parameters are omitted from the following discussion. Because of the higher turbine firing temperature and subsequently higher turbine exhaust temperature, steam conditions are 1,800 psi and 1,050 °F for the HP turbine and 405 psi and 1,050 °F for the IP turbine.

This two-train IGCC plant processes 6,200 tons per day of as-received coal to produce a net 688 MW of power. Overall efficiency is 38.8 percent (HHV basis). Improved performance over the Reference Case is demonstrated in Table 3-18.

Because of the higher turbine exit temperature in Case 7a, steam superheat temperature is 1,050 °F rather than 1,000 °F in Case 0. The increased coal flowrate, made possible by greater turbine throughput, leads to increased heat recovery in the gasifier, syngas quench, and flue gas through the HRSG – thus further contributing to increased steam turbine power generation in Case 7a.

The primary differences in auxiliary power consumption lay in the ASU main air compressor and the nitrogen compressor. The reduced ASU main air compressor power in Case 7a is due to integration between the gas turbine air compressor and the ASU, which reduces the fresh air feed through the main air compressor and therefore reduced power consumption. The increased N2 compressor power consumption in Case 7a is due to greater flowrate through the gas turbine in Case 7a than in Case 0 because of the larger turbine.

Overall, net plant efficiency increases by 3.4 percentage points going from the 7FA syngas turbine to the 2010-AST syngas turbine. Compared to Case 2a, the 2010-AST turbine increases process efficiency by 0.9 percentage points over the advanced "F" frame turbine in the reference plant.

Cost Analysis (Case 7a)

Capital and O&M costs are compared with Case 0 results in Table 3-19. The 2010-AST turbine has a higher power rating, which increases coal flowrate to the process, and therefore equipment sizes throughout the plant; this is reflected in the greater EPCC and TPC costs in Case 7a.

On a $/kW basis, the TPC of the 2010-AST plant decreases by 18 % because of increased power production. Not only is the turbine power output of Case 7a greater, but the process efficiency is 3.4 percentage points greater than the 7FA case. The primary reasons for this are air integration from the gas turbine to the ASU, the greater efficiency of the 2010-AST syngas turbine, and the increased steam cycle superheat temperature.

Comparing cost of electricity, the $0.0 782/kW-hr of Case 7a is 17 % less than the $0.0927/kW-hr of Case 0 because of (1) larger gas turbine, which increases the plant output and therefore decreases the capital cost on a $/kW basis, and (2) increased plant efficiency due to the higher pressure ratio and firing temperature of the 2010-AST syngas turbine compared to the 7FA turbine.

The reference plant with advanced "F" frame syngas turbine (Case 2a), by comparison, has $1,809/kW TPC and COE of $ 0.0814/kW-hr. The reference plant with 2010-AST turbine represents a 4 % reduction over the reference plant with advanced "F" frame turbine both in TPC (on a $/kW basis) and in COE.

3.8.2. Cumulative Impact of R&D

Composite Process Configuration (Case 7): Coal Feed Pump, Cryogenic ASU, Warm Gas Cleanup, 2010-AST Syngas Turbine, 85 % Capacity Factor

Table 3-18. Performance Impact of 2010-AST Syngas Turbine in the Reference Plant

	Case 0	Case 7a
	Reference plant with 7FA	Reference plant with 2010-AST
Gas Turbine Power (MWe)	384	500
Fuel Gas Expander (MWe)	6	8
Steam Turbine Power (MWe)	223	310
Total Power Produced (MWe)	614	818
Auxiliary Power Use (MWe)	-127	-130
Net Power (MWe)	487	688
As-Received Coal Feed (lb/hr)	402,581	519,515
Net Heat Rate (Btu/kW-hr)	9,649	8,806
Net Plant Efficiency (HHV)	35.4 %	38.8 %

Table 3-19. Case 7a: Capital and O&M Cost Comparison

	Case 0			Case 7a		
	Reference plant with 7FA			Reference plant with 2010-AST		
Capital Cost ($1,000)						
Plant Sections	**EPCC**	**TPC**	**TPC $/kW**	**EPCC**	**TPC**	**TPC $/kW**
1 Coal and Sorbent Handling	25,685	30,82 1	63	30,087	36,1 04	52
2 Coal and Sorbent Prep & Feed	39,472	48,980	101	46,880	58,1 74	85
3 Feedwater & Balance of Plant	28,606	35,077	72	31,242	38,240	56
4a Gasifier	184,371	236,212	485	217,947	279,210	405
4 b Air Separation Unit	153,591	168,950	347	171,740	188,914	275
5 a Gas Cleanup	93,441	112,389	231	112,102	134,83 7	196
5b CO_2 Removal & Compression	0	0	0	0	0	0
6 Gas Turbine	91,110	105,058	215	108,790	125,399	182
7 HRSG	44,560	49,511	102	52,175	57,924	84
8 Steam Cycle and Turbines	47,842	54,310	112	60,308	68,528	100
9 Cooling Water System	20,099	24.23 3	50	23,125	27,874	41
10 Waste Solids Handling System	34,981	38,752	80	40,939	45,347	66
11 Accessory Electric Plant	55,772	66,529	137	59,509	70,849	103
12 Instrumentation & Control	18,982	23,178	48	19,099	23,321	34
13 Site Preparation	13,956	18,143	37	14,350	18,655	27
14 Buildings and Structures	14,012	16,314	34	15,208	17,690	26
Total	866,482	1,028,457	2,113	1,003,501	1,191,067	1,731
O&M Cost ($1,000)						
Fixed Costs	**Total**		**% EPCC**	**Total**		**% EPCC**
Labor	19,542		2.26	22,548		2.25
Variable Operating Costs*	**Total**		**% EPCC**	**Total**		**% EPCC**
Maintenance Materials	18,368		2.12	22,390		2.23
Water	1,451		0.17	1,641		0.16
Chemicals	1,021		0.12	1,283		0.13
Waste Disposal	2,262		0.26	2,896		0.29
Total Variable Costs	23,102		2.67	28,210		2.81
Total O&M Cost	42,644		4.92	50,758		5.06
Fuel Cost*	55,690		6.43	71,865		7.16
Discounted Cash Flow Results						
Total Plant Cost ($/kW)			2,113			1,731
Levelized Cost of Electricity ($/kW-hr)			0.0927			0.0782

*Includes 75 % capacity factor

Table 3-20 below demonstrates improved overall process performance when the advanced "F" frame syngas turbine in Case 6 is replaced with a somewhat larger and more advanced 2010-AST turbine in Case 7.

Steam turbine power generation increases with the 2010-AST syngas turbine. Two reasons for this are increased coal feedrate and higher HRSG inlet temperature. The higher coal feedrate generates more heat in the gasifier and syngas cooling sections, which increases heat recovery for steam generation. The higher HRSG inlet temperature provides more sensible heat to the HRSG for steam generation.

Total auxiliary power consumption is very nearly identical between the two cases. The main air compressor in Case 7 consumes less power because of greater air extraction from the gas turbine, but the nitrogen compressor power consumption increases because more fuel, and therefore dilution nitrogen, is fed to the turbine.

The total power production and the net power production increase with the 2010-AST syngas turbine because of greater turbine power output and also more steam generation due to greater coal feedrate.

Overall, there is a 1.1 percentage point improvement – increasing net plant efficiency from 42.4 % to 43.5 %. This is a slightly greater improvement than in the reference plant comparison, in which performance improved by only 0.9 percentage point. The reason is the slurry feed gasifier in the reference plant; coal feedrate increases for the 2010-AST turbine, which increases the amount of slurry water to be evaporated in the reference plant – which is then condensed during cold gas cleanup and thus part of the energy contributed by increased coal flowrate in the slurry feed case is lost, resulting in reduced performance improvement than the coal feed pump case. This is one example of synergy resulting from advanced technologies.

Cost Analysis (Case 7)

The increased TPC of Case 7, shown in Table 3-21, reflects increased coal feedrate made possible by the larger gas turbine; increased coal feedrate increases size and throughput of all other process equipment. The corresponding increase in net power production however, when divided into the TPC, results in nearly a uniform 5 % reduction in TPC on a $/kW basis in all cost accounts. The TPC on a $/kW basis reduces from $ 1,588/kW to $1,516/kW. Because COE is dominated by capital cost, the COE reduces by 4 % from $0.0675/kW-hr to $0.0648/kW-hr.

Table 3-20. Incremental Performance Improvement from the 2010-AST Turbine

	Case 6	Case 7
	Coal pump, full WGCU, 85% CF, advanced "F"	Coal pump, full WGCU, 85%CF, 2010-AST
Gas Turbine Power (MWe)	464	500
Fuel Gas Expander (MWe)	9	9
Steam Turbine Power (MWe)	310	323
Total Power Produced (MWe)	783	832
Auxiliary Power Use (MWe)	-119	-118
Net Power (MWe)	664	714
As-Received Coal Feed (lb/hr)	457,710	480,583
Net Heat Rate (Btu/kW-hr)	8,042	7,849
Net Plant Efficiency (HHV)	42.4 %	43.5 %

These are approximately the same cost reductions over the advanced "F" frame syngas turbine as observed in the reference plant Case 7a.

Table 3-21. Case 7: Capital and O&M Cost Comparison

	Case 6			Case 7		
	Coal pump, full WGCU, 85% CF, advanced "F"			Coal pump, full WGCU, 85% CF, 2010-AST		
Capital Cost ($1,000)						
Plant Sections	**EPCC**	**TPC**	**TPC $/kW**	**EPCC**	**TPC**	**TPC $/kW**
1 Coal and Sorbent Handling	27,811	33,373	50	28,667	34,400	48
2 Coal and Sorbent Prep & Feed	44,617	55,096	83	46,109	56,939	80
3 Feedwater & Balance of Plant	27,451	33,552	51	27,961	34,165	48
4a Gasifier	193,316	247,855	373	199,598	255,906	358
4b Air Separation Unit	146,987	161,686	244	149,311	164,243	230
5a Gas Cleanup	68,801	82,937	125	71,294	85,943	120
5b CO_2 Removal & Compression	0	0	0	0	0	0
6 Gas Turbine	103,747	119,599	180	109,034	125,681	176
7 HRSG	51,262	56,927	86	51,973	57,703	81
8 Steam Cycle and Turbines	60,341	68,566	103	62,144	70,623	99
9 Cooling Water System	23,016	27,744	42	23,475	28,295	40
10 Waste Solids Handling System	35,793	39,651	60	36,888	40,863	57
11 Accessory Electric Plant	57,657	68,632	103	58,336	69,406	97
12 Instrumentation & Control	18,636	22,755	34	18,600	22,711	32
13 Site Preparation	14,063	18,282	28	14,127	18,366	26
14 Buildings and Structures	14,988	17,435	26	15,149	17,620	25
Total	888,486	1,054,090	1,588	912,666	1,082,864	1,516
O&M Cost ($1,000)						
Fixed Costs	**Total**		**% EPCC**	**Total**		**% EPCC**
Labor	21,045		2.37	21,045		2.31
Variable Operating Costs	**Total**		**% EPCC**	**Total**		**% EPCC**
Maintenance Materials	24,594		2.77	25,698		2.82
Water	1,343		0.15	1,374		0.15
Chemicals	4,808		0.54	5,048		0.55

Table 3-21. (Continued)

Variable Operating Costs	Total	% EPCC	Total	% EPCC
Waste Disposal	2,651	0.30	2,779	0.30
Total Variable Costs	33,397	3.76	34,899	3.82
Total O&M Cost	54,442	6.13	55,944	6.13
Fuel Cost	71,758	8.08	75,344	8.26
Discounted Cash Flow Results				
Total Plant Cost ($/kW)		1,588		1,516
Levelized Cost of Electricity ($/kW-hr)		0.0675		0.0648

3.9. Ion Transport Membrane

An alternative to cryogenic air separation, the ion transport membrane (ITM) produces a pure oxygen permeate stream at low pressure, leaving the nitrogen-rich non-permeate at high pressure for fuel stream dilution and expansion through the gas turbine. Although the oxygen must be compressed to gasifier pressure, the ITM has the advantage of reducing auxiliary power required to compress dilution nitrogen, and thus improves process efficiency.

The ITM is a ceramic perovskite-type material that, at high temperature (800-900 °C), allows the passage of oxygen ions across the ceramic membrane. Thes e ions re-combine to form oxygen molecules on the permeate side of the membrane. In this manner, oxygen is separated from air to produce 100 percent pure oxygen. A small amount of clean fuel is oxidized directly with the ITM air feed stream to heat the air stream to ITM temperature.

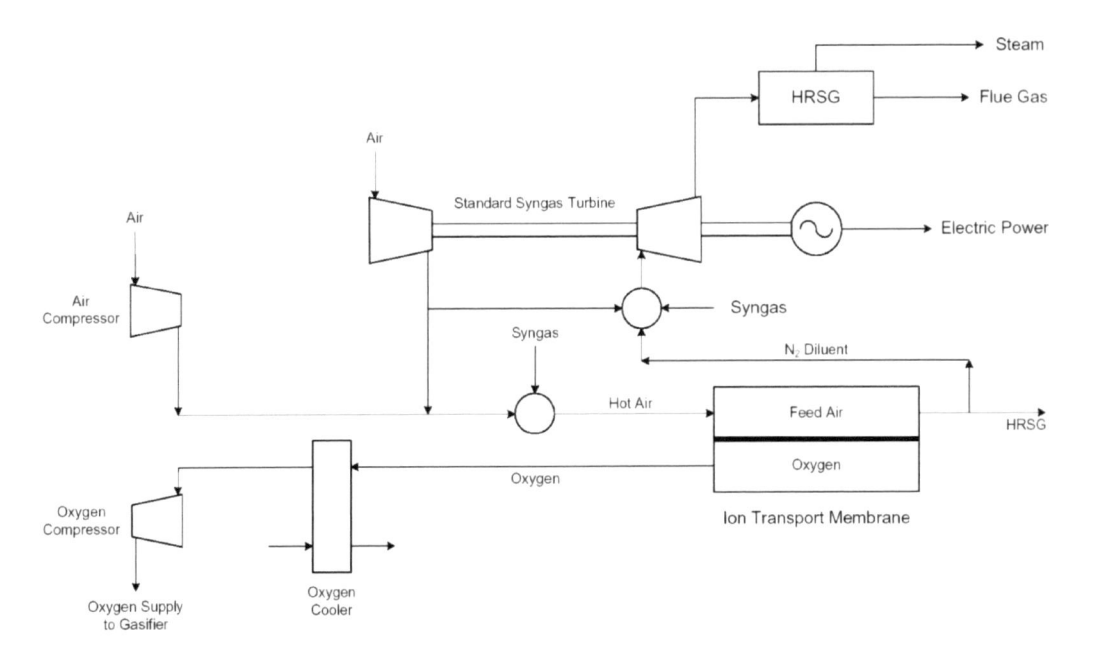

Figure 3-6. Partial ITM Air Integration Configuration.

A recent Gas Turbine World article [8] describes alternative configurations for full air integration, partial integration, and zero integration. The full air integration scenario includes a recuperator to transfer heat from the hot non-permeate stream to the feed air stream in order to reduce the amount of syngas used to heat the ITM. The recuperator increases capital equipment cost and also introduces a pressure drop across the ITM system that introduces the need for a boost compressor in order to provide sufficient pressure for the non-permeate stream to return to the turbine. The partial integration scenario, illustrated below in Figure 3-6, has no recuperator or boost compressor. Air extracted from the syngas urbine compressor is supplemented by a stand-alone air compressor to provide full air feed to the ITM. A small amount of syngas is oxidized to heat the feed air to the membrane. Oxygen is transported across the membrane; it must be cooled and compressed to gasifier pressure. With only a nominal pressure drop across the membrane, part of the nitrogen-rich non-permeate is used as syngas diluent in the turbine, and the remainder is expanded and cooled for heat recovery in the HRSG.

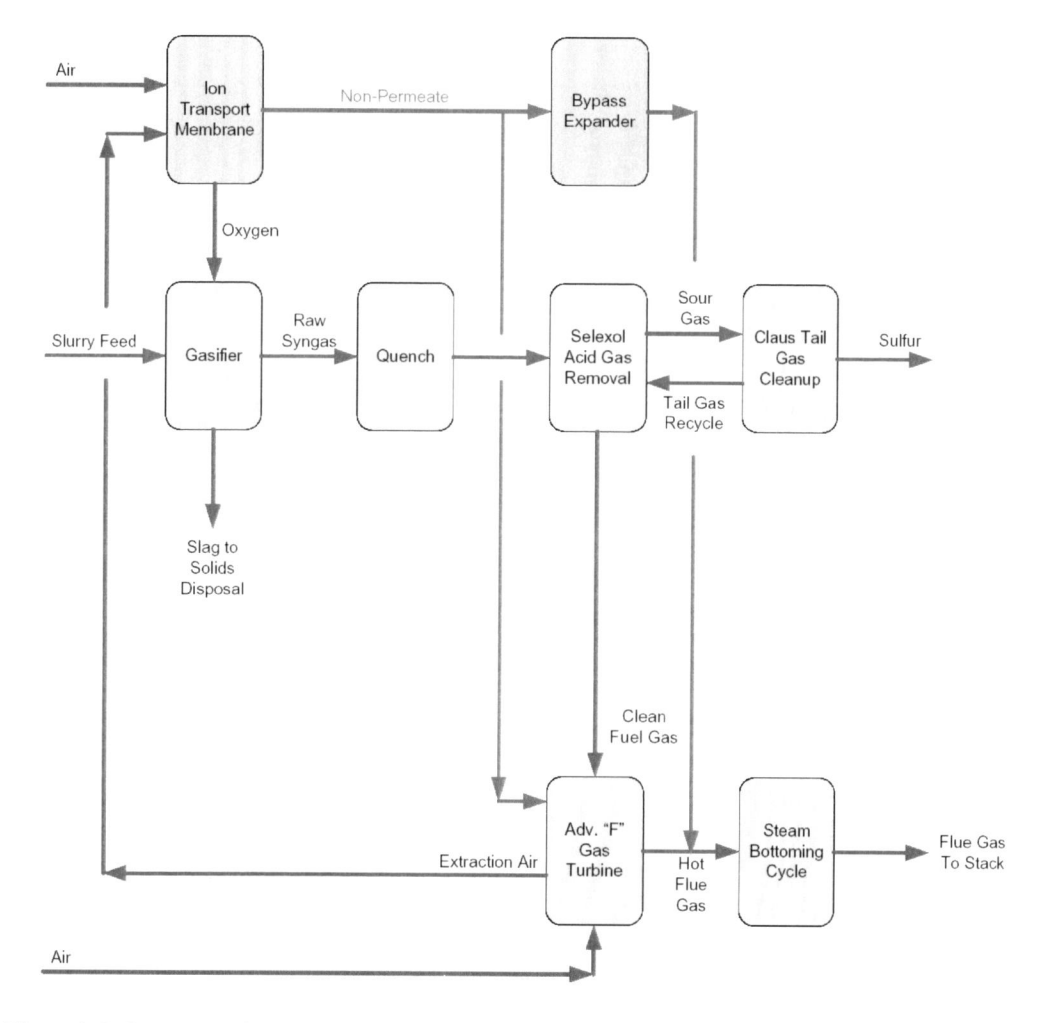

Figure 3-7. Case 8a: Reference IGCC Process With ITM Air Separation.

3.9.1. Impact of ITM in the Reference Plant (Case 8a)

Case 8a Configuration: Slurry Feed Gasifier, ITM, Cold Gas Cleanup, Advanced "F" Frame Syngas Turbine, 75 % Capacity Factor

A block flow diagram of the process with partial ITM air integration is shown in Figure 3-7; process details have been omitted to protect business-sensitive information. Note that this process is based on the advanced "F" frame syngas turbine for air extraction from the gas turbine; results from this case are compared against Case 2a, which is the reference process with advanced "F" frame syngas turbine. Although the Siemens SGT6-6000G turbine is the basis for the Gas Turbine World article, the advanced "F" frame turbine was chosen in this analysis for consistent comparison with other pathway study cases.

Table 3-22 below compares overall process performance improvement due to air separation using the ITM.

In the ITM case, part of the N_2-rich raffinate bypasses the gas turbine to produce 47 MW. Steam turbine power production increases by 23 MW due to eliminating the steam duty of the cryogenic ASU and also increased process heat recovery resulting from increased coal feed rate. Auxiliary power consumption increases by 28 MW as the result of increased air flowrate to the ASU and higher oxygen compression ratio; the total increase is partly offset by eliminating the nitrogen compressor. The incremental power production results in net plant efficiency increasing from 37.9 % to 38.2 % – an increase by 0.3 percentage points.

This increase in process efficiency is slightly lower than Air Products' expectations, although the values of net plant efficiency are different [9]; their baseline with cryogenic air separation had an efficiency of 38.4 % (HHV), increasing to 38.9 % with ITM air separation – an increase by 0.5 percentage points.

Table 3-22. Performance Impact of ITM in th Reference Plant

	Case 2a	Case 8a
	Reference plant with adv. "F" and cryogenic ASU	Reference plant with advanced "F" and ITM
Gas Turbine Power (MWe)	464	464
Fuel Gas Expander (MWe)	8	8
Bypass Expander (MWe)	NA	47
Steam Turbine Power (MWe)	293	316
Total Power Produced (MWe)	765	834
Auxiliary Power Use (MWe)	-128	-156
Net Power (MWe)	637	678
As-Received Coal Feed (lb/hr)	491,633	518,870
Net Heat Rate (Btu/kW-hr)	9,004	8,924
Net Plant Efficiency (HHV)	37.9 %	38.2 %

Cost Analysis (Case 8a)

Comparing capital costs between Cases 2a and 8a in Table 3-23, the TPC in most accounts such as coal handling, coal feed, BOP, gasifier, gas cleanup, HRSG, steam cycle, and waste solids handling system increase by between 2-8 percent because of higher plant throughput due to increased coal feed necessary to heat the ITM. The most significant difference occurs in the air separation unit (ITM) sub-account.

The bare erected cost of the ITM is assumed to be 67 % of the cost of an equivalent cryogenic ASU. Because of the different plant sizes, it is about 77 % of the cost of the cryogenic ASU in Case 2a. This reduces the cost of the ASU by about $42 MM, which equates to about $79/kW decrease in the cost of the ASU.

Table 3-23. Case 8a: Capital and O&M Cost Comparison

	Case 2a			Case 8a		
	Reference plant with advanced "F" and cryogenic ASU			Reference plant with Advanced "F" and ITM		
Capital Cost ($1,000)						
Plant Sections	**EPCC**	**TPC**	**TPC $/kW**	**EPCC**	**TPC**	**TPC $/kW**
1 Coal and Sorbent Handling	29,076	34,890	55	30,063	36,076	53
2 Coal and Sorbent Prep & Feed	45,169	56,050	88	46,839	58,124	86
3 Feedwater & Balance of Plant	30,636	37,513	59	31,227	38,223	56
4a Gasifier	210,196	269,284	423	217,002	278,016	410
4b Air Separation Unit	167,073	183,781	289	118,445	142,134	210
5a Gas Cleanup	107,769	129,625	203	110,816	133,289	197
5b CO$_2$ Removal & Compression	0	0	0	0	0	0
6 Gas Turbine	103,491	119,302	187	103,564	119,387	176
7 HRSG	50,936	56,565	89	52,626	58,460	86
8 Steam Cycle and Turbines	57,934	65,820	103	61,125	69,461	102
9 Cooling Water System	22,515	27,140	43	23,338	28,130	41
10 Waste Solids Handling System	39,568	43,829	69	40,909	45,314	67
11 Accessory Electric Plant	58,402	69,559	109	62,683	74,697	110
12 Instrumentation & Control	19,010	23,212	36	20,133	24,584	36
13 Site Preparation	14,247	18,522	29	14,554	18,920	28
14 Buildings and Structures	14,974	17,421	27	15,412	17,928	26
Total	970,995	1,152,513	1,809	948,736	1,142,740	1,685
O&M Cost ($1,000)						
Fixed Costs	**Total**		**% EPCC**	**Total**		**% EPCC**
Labor	22,548		2.32	22,548		2.38

Table 3-23. (Continued)

Variable Operating Costs*	Total	% EPCC	Total	% EPCC
Maintenance Materials	21,339	2.20	22,71 5	2.39
W ater	1,596	0.16	1,56 5	0. 17
Chemicals	1,215	0.13	1,298	0.14
Waste Disposal	2,745	0.28	2,893	0.31
Total Variable Costs	26,896	2.77	28,47 1	3.00
Total O&M Cost	49,444	5.09	51,02 0	5.38
F uel Cost*	68,008	7.00	71,77 6	7.57
Discounted Cash Flow Results				
Total Plant Cost ($/kW)		1,809		1,685
Levelized Cost of Electricity ($/kW- hr)		0.0814		0.0775

*Includes 75% capacity factor

Overall, the total plant cost decreases by $10 MM going from Case 2a to Case 8a. Because of increased net power production (678 MW vs. 637 MW) however, the TPC cost on a $/kW basis decreases from $1,809/kW to $1,685/kW. Annual fuel cost increases slightly due to additional coal needed to heat the ITM.

Despite the increased O&M and fuel costs, the decreased capital cost on a $/kW basis drives the COE down from $0.0814/kW-hr to $0.0775/kW-hr – a decrease of about 4.8 % in cost of electricity.

These cost and performance results are somewhat less than Air Products' expectations [10]. Air Products projects IGCC net power output to increase by 15 %; the increase from 637 MW to 678 MW is a 6 % increase. While the plant efficiency is expected to increase by 0.5 percentage point, the increase in Noblis' simulation is 0.3 percentage point. Air Products estimates the oxygen plant cost to decrease by 25 % on a $/sTPD O_2 basis; Noblis' estimate is a 23 % decrease on a $/kW basis (equal to a 27 % decrease on a $/sTPD basis). Finally, Air Products predicts the TPC on a $/kW basis to decrease by 9 %; Noblis' reduction from $1,809/kW to $1,685/kW represents a 6.9 % decrease. There are some proprietary aspects of Air Products' process that are not included in this analysis; this may explain the differences between Noblis and Air Products results.

3.9.2. Cumulative Impact of R&D

Composite Process Configuration (Case 8): Coal Feed Pump, ITM, Warm Gas Cleanup, 2010-AST Syngas Turbine, 85 % Capacity Factor

A block flow diagram of the ITM air separation unit implemented in the advanced technology process configuration is shown in Figure 3-8. Because the dry feed gasifier requires less oxygen than the slurry feed gasifier, the ITM will be smaller in this case; this represents a reduction in auxiliary power consumption compared to the reference process, and also eliminates the bypass expander. Table 3-24 below compares the overall process performance for the advanced technology case when the ITM replaces cryogenic ASU.

Table 3-24. Incremental Performance Improvement from ITM

	Case 7	Case 8
	Coal pump, full WGCU, 85% CF, 2010-AST, cryogenic ASU	Coal pump, full WGCU, 85% CF, 2010-AST, ITM
Gas Turbine Power (MWe)	500	500
Fuel Gas Expander (MWe)	9	9
Steam Turbine Power (MWe)	323	340
Total Power Produced (MWe)	832	849
Auxiliary Power Use (MWe)	-118	-123
Net Power (MWe)	714	725
As-Received Coal Feed (lb/hr)	480,583	480,947
Net Heat Rate (Btu/kW-hr)	7,849	7,734
Net Plant Efficiency (HHV)	43.5 %	44.1 %

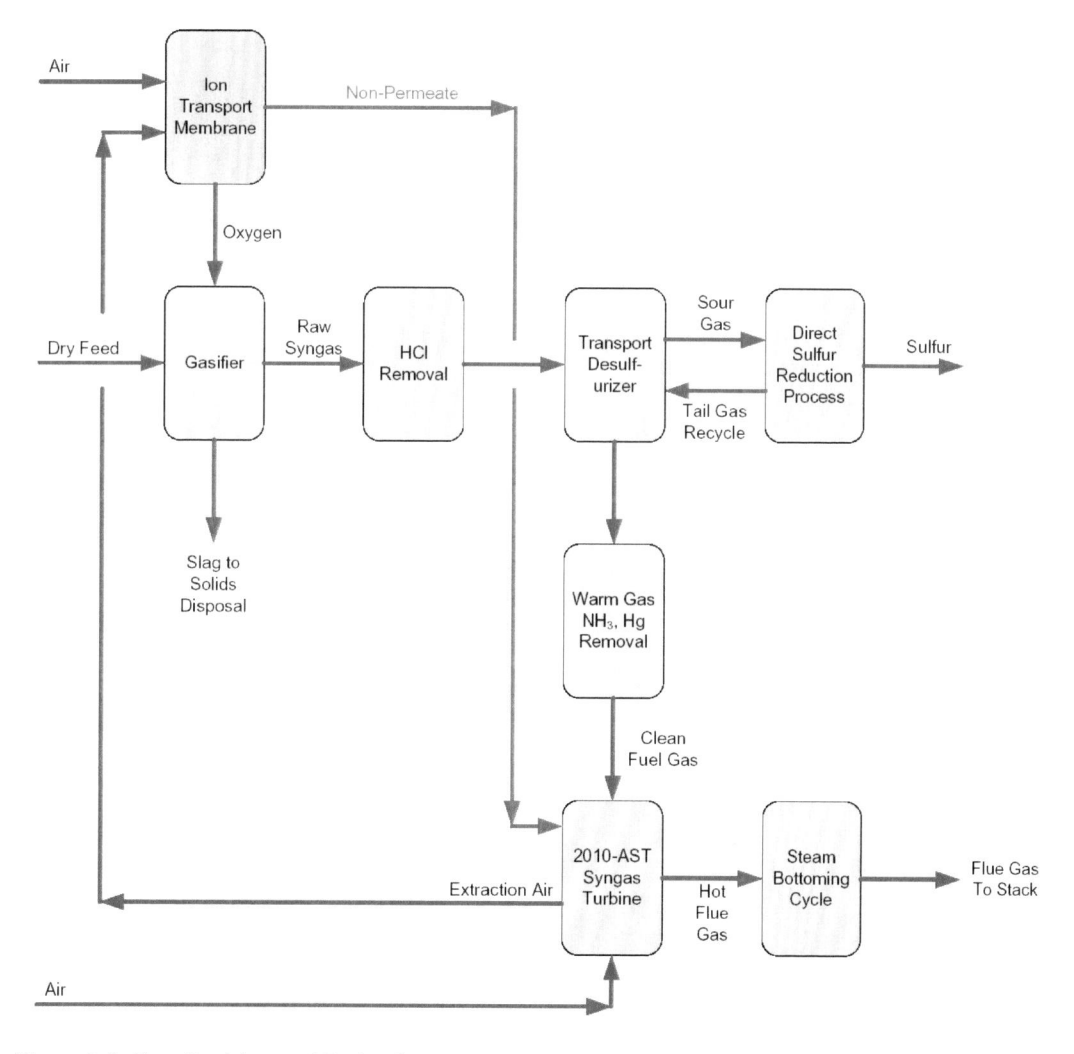

Figure 3-8. Case 8: Advanced Technology Process With ITM Air Separation.

The steam turbine power production in Case 8 increases by 17 MW because (1) the ITM air separation unit eliminates two significant steam requirements used in a cryogenic ASU; (2) steam added to humidify the fuel gas in the cryogenic process is replaced by water spray that is evaporated by cooling the N2-rich raffinate, and (3) heat recovery from the oxygen and fluff nitrogen coolers in the ITM process.

Auxiliary power use decreases by 5 MW in Case 8 – the result of tradeoffs between main air compressor, ITM boost compressor, and oxygen compressors vs. eliminating the nitrogen compressor.

With the same coal feed rate, net power production increases by 11 MW – improving net plant efficiency from 43.5 % to 44.1 %.

While the ITM increases process efficiency by 0.3 percentage points in the reference process with slurry feed gasifier, cold gas cleanup, and advanced "F" frame turbine (Cases 2a and 8a), it has a slightly better performance improvement with coal feed pump, warm gas cleanup, and 2010-AST turbine – improving process efficiency by 0.6 percentage points. The dry feed gasifier in Case 8 requires less oxygen than slurry feed, and reduces the N2-rich raffinate flowrate enough that all can be used as diluent in the gas turbine, eliminating the bypass expander. This increases the temperature at the inlet to the HRSG, thus increasing heat recovery in the steam cycle. The ample supply of diluent also eliminates the need for humidifying steam (as needed in Case 7).

Cost Analysis (Case 8)

Comparing capital costs between Cases 7 and 8 in Table 3-25, the TPC in most accounts such as coal handling, coal feed, BOP, gasifier, gas cleanup, HRSG, steam cycle, and waste solids handling system are very similar between cases; this is due to nearly the same coal feed rates. The only significant difference occurs in the air separation unit (ITM) sub-account.

Table 3-25. Case 8: Capital and O&M Cost Comparison

	Case 7			Case 8		
	Coal pump, full WGCU, 85% CF, 2010-AST, cryogenic ASU			Coal pump, full WGCU, 85% CF, 2010-AST, ITM		
Capital Cost ($1,000)						
Plant Sections	**EPCC**	**TPC**	**TPC $/kW**	**EPCC**	**TPC**	**TPC $/kW**
1 Coal and Sorbent Handling	28,667	34,400	48	28,682	34,418	47
2 Coal and Sorbent Prep & Feed	46,109	56,939	80	46,132	56,967	79
3 Feedwater & Balance of Plant	27,961	34,165	48	27,968	34,174	47
4a Gasifier	199,598	255,906	358	198,789	254,893	351
4b Air Separation Unit	149,311	164,243	230	100,238	120,285	166
5a Gas Cleanup	71,294	85,943	120	69,056	83,223	115

Table 3-25. (Continued)

Plant Sections	Case 7 Coal pump, full WGCU, 85% CF, 2010-AST, cryogenic ASU			Case 8 Coal pump, full WGCU, 85% CF, 2010-AST, ITM		
	EPCC	TPC	TPC $/kW	EPCC	TPC	TPC $/kW
5b CO$_2$ Removal & Compression	0	0	0	0	0	0
6 Gas Turbine	109,034	125,681	176	108,969	125,605	173
7 HRSG	51,973	57,703	81	51,988	57,719	80
8 Steam Cycle and Turbines	62,144	70,623	99	64,435	73,238	101
9 Cooling Water System	23,475	28,295	40	24,056	28,994	40
10 Waste Solids Handling System	36,888	40,863	57	36,901	40,877	56
11 Accessory Electric Plant	58,336	69,406	97	59,237	70,486	97
12 Instrumentation & Control	18,600	22,711	32	18,833	22,996	32
13 Site Preparation	14,127	18,366	26	14,177	18,430	25
14 Buildings and Structures	15,149	17,620	25	15,388	17,896	25
Total	912,666	1,082,864	1,516	864,848	1,040,201	1,434

O&M Cost ($1,000)

Fixed Costs	Total		% EPCC	Total		% EPCC
Labor	21,045		2.31	19,542		2.26
Variable Operating Costs	**Total**		**% EPCC**	**Total**		**% EPCC**
Maintenance Materials	25,698		2.82	26,066		3.01
Water	1,374		0.15	1, 314		0.15
Chemicals	5,04 8		0.55	5,060		0. 59
Waste Disposal	2,779		0.30	2,781		0.32
Total Variable Costs	34,899		3.82	35,220		4.07
Total O&M Cost	55,944		6.13	54,76 2		6.33
F uel Cost	75,344		8.26	75,401		8.72

Discounted Cash Flow Results

Total Plant Cost ($/kW)	1,516	1,434
Levelized Cost of Electricity ($/kW-hr)	0.0648	0.0622

The bare erected cost of the ITM is assumed to be 67 % of the cost of an equivalent cryogenic ASU; with the slight difference in gasifier throughput, it is about 73 % of the cost of the cryogenic ASU of Case 7. This reduces the cost of the ASU by about $44 MM, which equates to about $64/kW decrease in the cost of the ASU.

Overall, the total plant cost decreases by $43 MM going from Case 7 to Case 8. Because of slightly increased net power production (725 MW vs. 714 MW), the TPC cost on a $/kW basis decreases from $1,516/kW to $1,434/kW.

With nearly equal O&M and fuel costs, the decreased capital cost drives the COE down from $0.0648/kW-hr to $0.0622/kW-hr – a decrease of about 4.0 % in cost of electricity.

3.10. Advanced Syngas Turbine – 2015-AST

By 2015, DOE has more aggressive performance efficiency improvement targets for advanced syngas turbine technology. Superior to the 2010-AST turbine, the 2015-AST performance improvements are expected from increased pressure ratio and turbine inlet temperature, further improving efficiency of the gas turbine.

3.10.1. Impact of 2015-A ST Syngas Turbine in the Reference Plant (Case 9a)

Case 9a Configuration: Slurry Feed Gasifier, Cryogenic ASU, Cold Gas Cleanup, 2015-AST Syngas Turbine, 75 % Capacity Factor

The process block flow diagram of the reference IGCC process with a 2015-AST syngas turbine is identical to Figure 3-2 from Case 2a. Like the advanced "F" frame and 2010-AST syngas turbines, the 2015-AST produces more power, has a higher pressure ratio, and higher firing temperature than the 7FA syngas turbine. Turbine performance parameters for the 201 5-AST are omitted from the following discussion in order to protect business-sensitive information. The increased turbine exhaust temperature (over that of the 7FA) enables steam superheat and reheat temperatures to 1,050 °F.

A single process train IGCC plant processes 4,300 tons per day of as-received coal to produce a net 500 MW of power. Overall efficiency is 40.5 percent (HHV basis). Carbon utilization is 98 percent and the capacity factor is 75 percent. Total power generated includes 6 MW from the fuel gas expander and 213 MW from the steam turbine. Auxiliary power use is estimated to be 89 MW. Performance improvement resulting from the 201 5-AST is compared to the Reference Case in Table 3-26.

Table 3-26. Performance Impact of 2015-AST Syngas Turbine in the Reference Plant

	Case 0	Case 9a
	Reference plant with 7FA	Reference plant with 2015-AST
Gas Turbine Power (MWe)	384	370
Fuel Gas Expander (MWe)	6	6
Steam Turbine Power (MWe)	223	213
Total Power Produced (MWe)	614	589
Auxiliary Power Use (MWe)	-127	-89
Net Power (MWe)	487	500
As-Received Coal Feed (lb/hr)	402,581	361,531
Net Heat Rate (Btu/kW-hr)	9,649	8,435
Net Plant Efficiency (HHV)	35.4 %	40.5 %

Coal flowrate decreases in Case 9a due to combined reduced gas turbine power output and improved gas turbine efficiency. The reduced coal flowrate also reduces heat recovery from the gasifier and syngas cooling, leading to decreased steam turbine power generation.

The primary difference in auxiliary power consumption is the ASU main air compressor due to integration between the gas turbine air compressor and the ASU, which reduces the fresh air feed through the main air compressor and therefore reduced power consumption. All other auxiliary power accounts in Case 9a are generally less than Case 0 due to decreased coal flowrate and therefore decreased throughput by all plant sections.

Overall, the net plant efficiency increases by 5.1 percentage points going from the 7FA syngas turbine to the 2015-AST syngas turbine.

Cost Analysis (Case 9a)

Capital and O&M costs are compared with Case 0 results in Table 3-27. The choice of gas turbine is the reason for differences in capital costs between Case 0 (7FA turbine) and Case 9a (2015-AST turbine). The 2015-AST turbine has a higher power rating per unit and the number of turbine trains reduces from two to one in Case 9a. The reduction in number of trains – essentially providing one large train of gasification, ASU, gas cleanup, and gas turbine rather than two of each in Case 0 – represents an economy of scale, which explains the decreased cost of the gasifier, ASU, gas cleanup, and gas turbine sections in Case 9a.

On a $/kW basis, the TPC of the 2015-AST plant also decreases. Not only is the turbine power output of Case 9a greater, but the process efficiency is 5.1 percentage points greater than the 7FA case. As described above, the primary reasons for this are air integration from the gas turbine to the ASU, the greater efficiency of the 2015-AST syngas turbine, and the increased steam cycle superheat temperature.

Comparing cost of electricity, the $0.0768/kW-hr of Case 9a is less than the $0.0927/kW-hr of Case 0 because of (1) larger gas turbine machine, which decreases the capital cost on a $/kW basis, and (2) increased plant efficiency due to the higher pressure ratio and firing temperature of the 2015-AST syngas turbine compared to the 7FA turbine.

Table 3-27. Case 9a: Capital and O&M Cost Comparison

Plant Sections	Case 0 Reference plant with 7FA			Case 9a Reference plant with 2015-AST		
	EPCC	TPC	TPC $/kW	EPCC	TPC	TPC $/kW
Capital Cost ($1,000)						
1 Coal and Sorbent Handling	25,685	30,821	63	24,025	28,829	58
2 Coal and Sorbent Prep & Feed	39,472	48,980	101	36,711	45,553	91
3 Feedwater & Balance of Plant	28,606	35,077	72	27,614	33,887	68
4a Gasifier	184,371	236,212	485	136,501	17 4,821	350
4b Air Separation Unit	153,591	168,950	347	109,034	119,938	240

Table 3-27. (Continued)

	Case 0			Case 9a		
	Reference plant with 7FA			Reference plant with 201 5-AST		
5a Gas Cleanup	93,441	112,389	231	69,903	84,086	168
5b CO$_2$ Removal & Compression	0	0	0	0	0	0
6 Gas Turbine	91,110	105,058	215	74,293	85,713	171
7HRSG	44,560	49,511	102	38,761	43,0 20	86
8 Steam Cycle and Turbines	47,842	54,3 10	112	46,184	52,4 24	105
9 Cooling Water System	20,099	24.2 33	50	19,359	23 ,346	47
10 Waste Solids Handling System	34,981	38,752	80	32,738	36,2 69	73
11 Accessory Electric Plant	55,772	66,529	137	50,334	59,9 38	120
1 2 Instrumentation & Control	18,982	23,17 8	48	17,152	20,9 44	42
13 Site Preparation	13,956	18,1 43	37	13,430	17,4 59	35
14 Buildings and Structures	14,012	16,314	34	13,621	15,85 8	32
Total	866,482	1,028,457	2,113	709,661	842,08 4	1,684

O&M Cost ($1,000)

Fixed Costs	**Total**		**% EPCC**	**Total**		**% EPCC**
Labor	19,542		2.26	16,535		2.33
Variable Operating Costs*	**Total**		**% EPCC**	**Total**		**% EPCC**
Maintenance Materials	18,368		2.12	17,880		2.52
Water	1,451		0.17	1,233		0.17
Chemicals	1,021		0.12	975		0.14
Waste Disposal	2,262		0.26	2,040		0.29
Total Variable Costs	23,102		2.67	22,128		3.12
Total O&M Cost	42,644		4.92	38,663		5.45
Fuel Cost*	55,690		6.43	50,011		7.05

Discounted Cash Flow Results

Total Plant Cost ($/kW)	2,113	1,684
Levelized Cost of Electricity ($/kW-hr)	0.0927	0.0768

*Includes 75 % capacity factor.

The reference process with advanced "F" frame syngas turbine (Case 2a), by comparison, has $1,809/kW TPC and COE of $0.0814/kW-hr. The reference process with 2010-AST syngas turbine (Case 7a) has $1,731/kW TPC and COE of $0.0782/kW-hr. The TPC of the 2015-AST syngas turbine process is the least due to capital cost savings resulting from the decrease of two trains to one, and the COE benefits from a reduction in number of plant operators, decreased O&M cost, and decreased fuel cost.

3.10.2. Cumulative Impact of R&D

Composite Process Configuration (Case 9): Coal Feed Pump, ITM, Warm Gas Cleanup, 2015-AST Syngas Turbine, 85 % Capacity Factor

Case 9 substitutes the 2015-AST syngas turbine for the 2010-AST syngas turbine that was present in Case 8. The table below compares the overall process performance for each plant.

Although the 2015-AST turbine produces more power per unit, there is only one train in Case 9 as opposed to two 2010-AST turbines in Case 8; turbine power consequently decreases, as does coal feedrate and all power accounts in general.

The 2015-AST syngas turbine increases process performance by 2.0 percentage points over Case 8. When the 2015-AST turbine replaces the 2010-AST turbine in the reference process, performance increases by 1.7 percentage points (Case 7a vs. Case 9a). In the cryogenic cases, the overpressure (the difference between turbine compressor delivery pressure and the ASU pressure) is lost – and more of it is lost in the 2015-AST case because of its higher pressure.

Operating at higher pressure than the gas turbine compressor, the ITM has a better "pressure match" with gas turbine compressor delivery pressure, which improves process efficiency and therefore the increased improvement of the 2015-AST over the 2010-AST in the ITM cases.

Cost Analysis (Case 9)

Comparing capital costs between Cases 8 and 9 in Table 3-29, the TPC in all accounts decreases because of reduced net power production, which corresponds to decreased coal flowrate and decreased plant equipment size, and therefore cost. The number of process trains (consisting of gasifier, ASU, gas cleanup, and gas turbine) decreases from two to one.

Table 3-28. Incremental Performance Improvement from the 2015-AST Turbine

	Case 8	Case 9
	Coal pump, 85% CF, full WGCU, ITM, 2010-AST	Coal pump, 85% CF, full WGCU, ITM, 2015-AST
Gas Turbine Power (MWe)	500	370
Fuel Gas Expander (MWe)	9	6
Steam Turbine Power (MWe)	340	228
Total Power Produced (MWe)	849	604
Auxiliary Power Use (MWe)	-123	-76
Net Power (MWe)	725	528
As-Received Coal Feed (lb/hr)	480,947	335,026
Net Heat Rate (Btu/kW-hr)	7,734	7,400
Net Plant Efficiency (HHV)	44.1 %	46.1 %

Table 3-29. Case 9: Capital and O&M Cost Comparison

	Case 8			Case 9		
	Coal pump, 85% CF, full WGCU, ITM, 2010-AST			Coal pump, 85% CF, full WGCU, ITM, 2015-AST		
Capital Cost ($1,000)						
Plant Sections	**EPCC**	**TPC**	**TPC $/kW**	**EPCC**	**TPC**	**TPC $/kW**
1 Coal and Sorbent Handling	28,682	34,418	47	22,918	27,502	52
2 Coal and Sorbent Prep & Feed	46,132	56,967	79	36,148	44,636	85
3 Feedwater & Balance of Plant	27,968	34,174	47	24,521	30,037	57
4a Gasifier	198,789	254,893	351	124,758	159,909	303
4b Air Separation Unit	100,238	120,285	166	63,491	76,189	144
5a Gas Cleanup	69,056	83,223	115	44,204	53,276	101
5b CO_2 Removal & Compression	0	0	0	0	0	0
6 Gas Turbine	108,969	125,605	173	74,457	85,902	163
7 HRSG	51,988	57,719	80	38,622	42,867	81
8 Steam Cycle and Turbines	64,435	73,238	101	48,479	55,039	104
9 Cooling Water System	24,056	28,994	40	19,870	23,960	45
10 Waste Solids Handling System	36,901	40,877	56	29,529	32,717	62
11 Accessory Electric Plant	59,237	70,486	97	48,666	57,891	110
12 Instrumentation & Control	18,833	22,996	32	16,427	20,058	38
13 Site Preparation	14,177	18,430	25	13,177	17,130	32
14 Buildings and Structures	15,388	17,896	25	13,580	15,809	30
Total	864,848	1,040,201	1,434	618,847	742,921	1,407
O&M Cost ($1,000)						
Fixed Costs	**Total**		**% EPCC**	**Total**		**% EPCC**
Labor	19,542		2.26	15,032		2.43
Variable Operating Costs	**Total**		**% EPCC**	**Total**		**% EPCC**
Maintenance Materials	26,066		3.01	20,611		3.33
Water	1,314		0.15	1,004		0.16
Chemicals	5,060		0.59	3,611		0.58
Waste Disposal	2,781		0.32	1,964		0.32
Total Variable Costs	35,220		4.07	27,191		4.39
Total O&M Cost	54,762		6.33	42,223		6.82
Fuel Cost	75,401		8.72	52,524		8.49
Discounted Cash Flow Results						
Total Plant Cost ($/kW)			1,434			1,407
Levelized Cost of Electricity ($/kW-hr)			0. 0622			0.0615

The reduction in TPC represents a reverse economy of scale, as TPC on a $/kW basis increases in every account except for the gasifier, ASU, gas cleanup, and gas turbine sections. Those accounts reduce in cost because of the reduction from two trains to one large train. In all other accounts, although the TPC decreases the cost on $/kW increases because of the reduced net power production.

Overall, the total plant cost decreases by $297 MM going from Case 8 to Case 9, but because of the decreased power production, the cost on a $/kW basis decreases by only $27/kW or 1.9 %.

The number of laborers, and therefore the fixed O&M cost, decreases as the result of reduc ing from two process trains to a single train. Variable O&M and fuel cost also decrease significantly as the result of reduced plant output. There is a slight net decrease in COE from $0.0622/kW-hr to $0.0615/kW-hr – a 1.1 % decrease as the result of the 2015-AST syngas turbine.

3.11. Increased Capacity Factor to 90 Percent

In Case 10, the process configuration remains the same as Case 9 (with process performance remaining the same as in Table 3-28 above for Case 9), but the capacity factor increases from 85 percent to 90 percent. This increased on-stream factor reflects anticipated improvements in process reliability, availability, and maintainability (RAM) due to DOE-sponsored R&D (with no additional capital or fixed O&M cost).

The differences between Case 9 and Case 10 lie in variable O&M costs, fuel cost, and plant revenues as the result of longer hours of operation. Variable O&M costs increase by about $1.6 MM/year, and fuel costs increase by about $3.1 MM/year. The increased plant revenue from additional power production results in decreased cost of electricity from $0.0615/kW-hr in Case 9 to $0.0595/kW-hr in Case 10 – a savings of about 3.3 % in cost of electricity resulting from increased capacity factor.

3.12. Pressurized Solid Oxide Fuel Cell

The solid oxide fuel cell offers potential for high efficiency conversion of chemical potential into electrical energy. Because the overall reaction of syngas and oxygen to form CO_2 and H_2O is exothermic, the fuel cell depends as much as possible on endothermic internal reforming of CH_4 to hydrogen and CO in order to limit temperature rise inside the fuel cell stack. Both the gasifier and fuel cell rely on a significant amount of steam, so warm gas cleanup is beneficial to avoid moisture condensation during desulfurization.

Process Configuration (Case 11): Catalytic Gasifier, Cryogenic ASU, Warm Gas Cleanup, Solid Oxide Fuel Cell, 90 % Capacity Factor

The fuel cell process configuration is based on a design proposed by SAIC [11]; the original process operating conditions were adopted in this study, and no further systems analysis attempt was made to optimize plant performance. This process feeds coal to a dry-feed, fluid bed, oxygen-blown catalytic gasifier. A block flow diagram is presented in Figure

3-9. Table 3-30 presents the calculated raw syngas composition, with high (16.6 mole percent) CH_4 content to promote reforming in the fuel cell.

Raw syngas passes through a high-efficiency cyclone to separate the bulk of entrained ash. The syngas then passes through a barrier filter (using ceramic or metal filter elements). Ash drained from the fluid bed gasifier and from the syngas cyclone is treated to separate and reprocess the gasification catalyst. The catalyst material is circulated back to the catalytic gasifier. A convective cooler generates steam, cooling the raw syngas from 1,300 °F to 950 °F.

The warm gas cleanup section is nearly identical to that used in the IGCC cases with GE gasifier, except that this case also includes a sulfur polishing step in order to attain very low sulfur concentration in the fuel cell feed stream. Raw syngas enters the chloride guard bed for HCl removal. The syngas is cooled in preparation for contact with zinc oxide sorbent, which reacts with H_2S to remove it from the syngas. To regenerate the sorbent, the ZnS transfers to the regenerator where it contacts with air and is oxidized. The SO_2 that is generated flows to the DSRP for sulfur recovery, and the regenerated sorbent is returned to the transport desulfurizer.

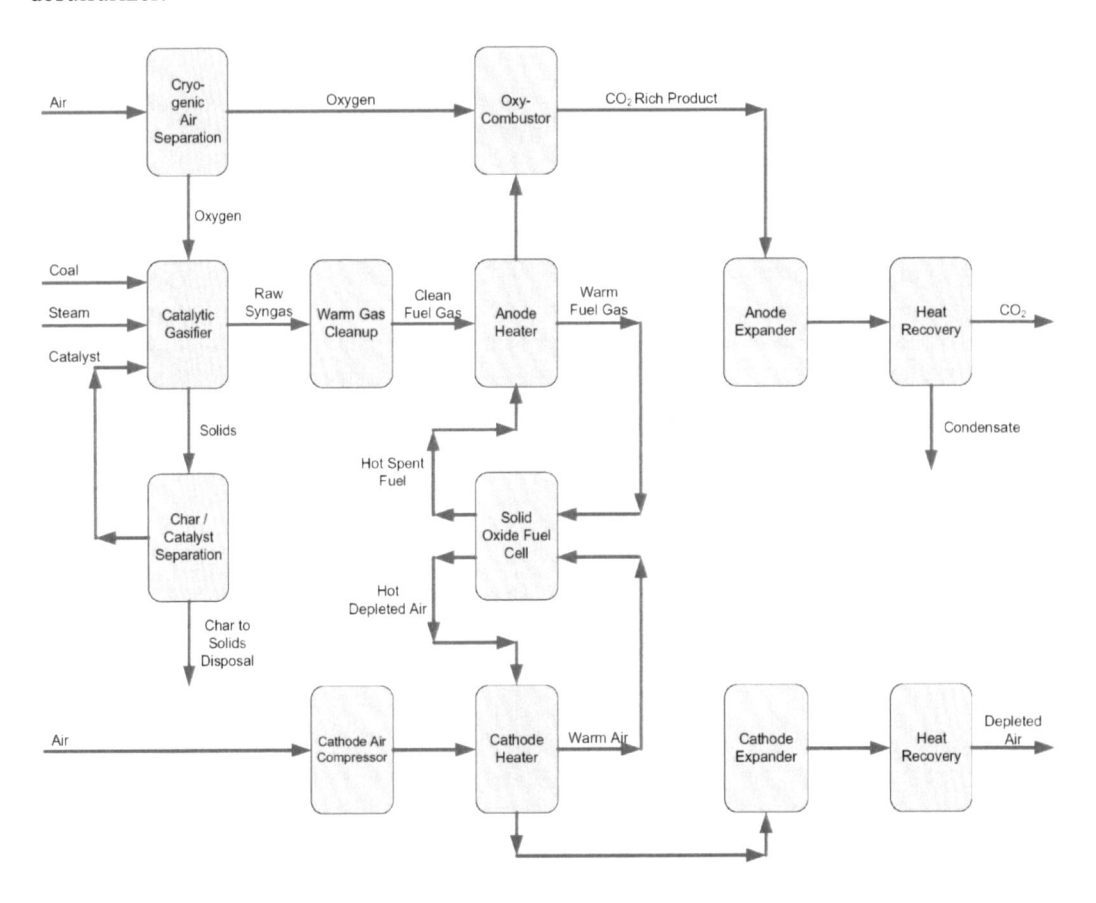

Figure 3-9. Case 11: Pressurized Solid Oxide Fuel Cell Process.

Table 3-30. Raw Syngas Composition from the Catalytic Gasifier

	Syngas Composition
H_2 (mole %)	15.0
CH_4	16.6
CO	4.7
CO_2	20.7
H_2O	41.8
N_2	0.4
H_2S	0.6
NH_3	626 ppm
Ar	192 ppm
HCl	689 ppm
COS	98 ppm

Desulfurized syngas, coming from the transport desulfurizer at 900 °F, passes through a polishing bed to reduce H2S concentration to a very low level. The desulfurized syngas is cooled to about 460 °F in preparation for mercury removal.

A small portion of the clean fuel gas exiting mercury removal is used as reducing gas in the DSRP. Here, regeneration gas from the Transport Desulfurization section is reduced, forming elemental sulfur:

$$SO_2 + 2\,H_2 = 2\,H_2O + S$$

$$3\,SO_2 + 2\,CH_4 = 2\,CO + 4\,H_2O + S$$

DSRP tail gas, containing H_2O and CO, is compressed and recycled to the transport desulfurizer. Elemental sulfur is condensed and removed as product.

Following chloride removal, desulfurization, and mercury removal, the clean fuel gas is ready for conversion in the fuel cell. High pressure steam can be added if necessary to adjust the hydrogen:carbon ratio in the fuel cell. The mixture is expanded to the fuel cell operating pressure of 275 psia. Heat exchange with the anode spent fuel stream heats the anode feed stream to 1,112 °F (600 °C).

On the cathode side, air is compressed to 290 psia. It is heated to 1,112 °F (600 °C) by the depleted air stream exiting the fuel cell cathode. Within the fuel cell anode, methane is completely reformed to CO and H_2. Oxygen diffuses from the cathode through the electrolyte to the anode, creating an electric potential, and reacts with H_2 that is formed from the equilibrium mixture of anode gases. Fuel conversion is assumed to be 85 %. The fuel cell inverter efficiency is assumed to be 96 %. A temperature rise of 150 °C from entrance to exit of the fuel cell is allowed; a large amount of cathode air must be circulated through the fuel cell to regulate temperature rise.

After heating the incoming cathode air stream, the spent cathode air enters an expander, and any remaining heat from the cathode expander exhaust is recovered for boiler feedwater heating. The spent anode fuel enters an oxy-combustor where remaining fuel is converted to flue gas. The hot flue gas is expanded and heat is recovered for steam generation. Following flue gas cooling, water is condensed and the remaining flue gas, consisting almost e ntirely of CO_2, can be compressed and transported for storage if sequestration is required.

All available process heat is collected to generate steam for the gasifier and the fuel cell anode; the process has no bottoming cycle. Table 3-31 below summarizes overall process performance.

Auxiliary power consumption is dominated by the large amount of cathode air that must be compressed and fed to the fuel cell in order to remove the large amount of heat generated from converting 85 % of the fuel gas in the fuel cell. The pressure is recovered in an expander, delivering 208 MW as part of the net power output.

Cos t Analysis (Case 11)

Table 3-32 summarizes the total plant cost, O&M cost, and fuel cost of the process. A fuel cell system TPC of $550/kW was assumed.[3] The fuel cell system includes fuel cell stack, anode and cathode heaters, anode steam generator and reheat, syngas expander, cathode air compressor, anode and cathode expanders, inverter, catalytic oxidizer and oxygen boost compressor, condensate knockout, and foundations. TPC includes equipment, labor, EPC services, and process and project contingencies. The same process and project contingencies used for the GE gasifier in the IGCC cases were assigned to the catalytic gasifier for this case. Ten percent process and ten percent project contingencies were assumed for the fuel cell system; this implies n^{th} plant design to evaluate the ultimate potential of fuel cell technology.

Total plant cost is based on a single process train. The total plant cost of $926 MM is less than the reference IGCC Case 0, but the fuel cell process generates nearly the same power as Case 3 with coal feed pump, cryogenic ASU, cold gas cleanup, and advanced "F" frame turbine. The large amount of power generated results in a TPC of $1,536/kW.

The very high process efficiency of 58.8 % results in a large cost savings in fuel. Particularly as fuel prices continue to put pressure on energy use, this high process efficiency will benefit fuel cell technology. The combination of low capital and fuel cost contributes to a COE of $0.0639/kW-hr (based on January 2007 dollars and 90 % capacity factor).

Table 3-31. Performance Summary of the SOFC Process

	Case 11
	SOFC
Fuel Cell Power (MW)	517
Syngas Expander (MW)	22
Cathode Air Expander (MW)	208
Anode Exhaust Expander (MW)	132
Total Power Produced (MW)	879
Auxiliary Power (MW)	-276
Net Power (MW)	603
As-Received Coal Feed (lb/hr)	300,000
Net Heat Rate (Btu/kW-hr)	5,805
Net Plant Efficiency (HHV)	58.8 %
Gasifier Cold Gas Efficiency	92.0 %

The fuel cell process configuration of Case 11 was developed by SAIC, and is described in NETL's report titled "The Benefits of SOFC for Coal-Based Power Generation" [11]. A comparison is provided in Appendix A.3 to support the results presented in Table 3-32.

Table 3-32. Case 11: Capital and O&M Cost Summary

	Case 11		
	Solid Oxide Fuel Cell		
Capital Cost ($1,000)			
Plant Sections	**EPCC**	**TPC**	**TPC $/kW**
1 Coal and Catalyst Handling	25,678	30,814	51
2 Coal and Catalyst Prep & Feed	33,550	41,428	69
3 Feedwater & Balance of Plant	17,805	21,649	36
4a Gasifier	123,853	155,335	258
4b Air Separation Unit	73,915	81,306	135
5a Gas Cleanup	55,056	66,351	110
5b CO2 Removal & Compression	0	0	0
6 Gas Turbine	0	0	0
7 Fuel Cell	278,195	333,836	554
8 Steam Cycle and Turbines	0	0	0
9 Cooling Water System	11,507	13,935	23
10 Waste Solids Handling System	32,217	35,692	59
11 Accessory Electric Plant	73,623	87,996	146
12 Instrumentation & Control	23,674	28,907	48
13 Site Preparation	14,480	18,823	31
14 Buildings and Structures	8,594	10,097	17
Total	772,147	926,169	1,536
O&M Cost ($1,000)			
Fixed Costs	**Total**		**% EPCC**
Labor	18,039		2.34
Variable Operating Costs*	**Total**		**% EPCC**
Maintenance Materials	28,316		3.67
Water	158		0.02
Chemicals	3,836		0.50
Fuel Cell Stack Replacement	17,835		2.31
Waste Disposal	2,397		0.31
Total Variable Costs	52,542		6.81
Total O&M Cost	70,581		9.14
Fuel Cost*	49,799		6.45
Discounted Cash Flow Results			
Total Plant Cost ($/kW)			1,536
Levelized Cost of Electricity ($/kW-hr)			0.0639

*Includes 90% capacity factor

4. SUMMARY OF ADVANCED TECHNOLOGY IMPROVEMENTS

The information presented in the previous section is consolidated in the following discussion in order to summarize the relative benefits of the advanced technologies that were investigated.

4.1. Impact of Individual Technologies

Process Efficiency

Figure 4-1 illustrates the performance improvement as each of the advanced technologies is evaluated individually within the reference plant. Because it represents multiple advanced technologies, the SOFC process is not included in this comparison.

The advanced syngas turbines provide the greatest performance improvements as the result of air integration, increased turbine firing temperature and pressure ratio, and increased HRSG inlet temperature. Compared to the 7FA turbine, the advanced "F" frame turbine improves process efficiency by 2.5 percentage points, the 2010-AST improves by 3.4 percentage points, and the 2015-AST improves by 5.1 percentage points.

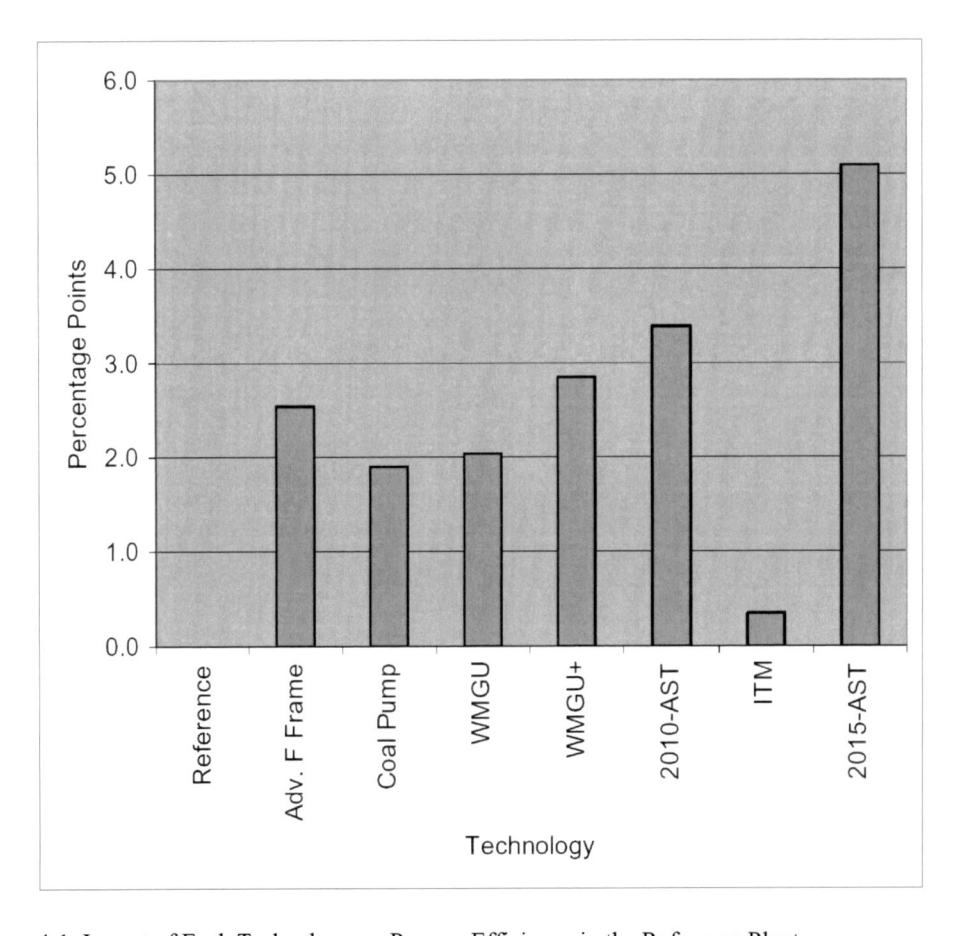

Figure 4-1. Impact of Each Technology on Process Efficiency in the Reference Plant.

Partial warm gas cleanup (WGCU) improves process efficiency by 2.0 percentage points as the result of eliminating sour water stripper and Selexol reboilers; full warm gas cleanup (WGCU+) adds another 0.8 percentage points to process efficiency by eliminating fuel gas reheat and eliminating loss of latent heat due to condensation during syngas quench.

The coal feed pump improves process efficiency by 1.9 percentage points by eliminating evaporation of slurry water in the gasifier, thus increasing the cold gas efficiency of the gasifi er (from 75.8 % to 81.6 %).

The ITM improves process efficiency by 0.3 percentage points (over the "reference" proc ess with advanced "F" frame syngas turbine). This is not as dramatic a performance improvement as the other technologies described above, but as described below the ITM's greater contribution will be to reduce TPC and COE resulting from lower capital cost than cryogenic air separation.

Total Plant Cost

Reductions in total plant cost (on a $/kW basis) are illustrated in Figure 4-2 as each techno logy is individually substituted into the reference plant. All costs are based on January 2007 dollars. The advanced "F" frame, 2010-AST, and 2015-AST syngas turbines result in the most significant capital cost reductions of all technologies (by $304/kW, $3 82/kW, and $429/kW, respectively). These reductions are due more to the increased net power generated than from any change in turbine equipment cost. The turbine section itself contributes only $28/kW, $32/kW, and $44/kW reduction to the total plant cost, respectively.

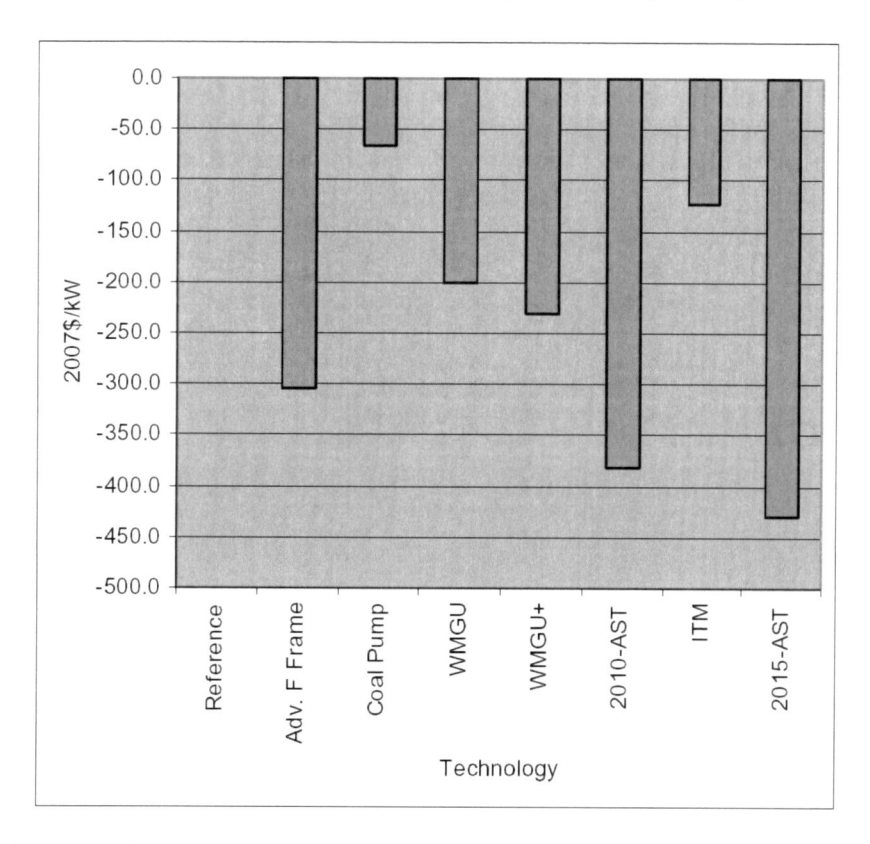

Figure 4-2. Impact of Each Technology on Total Plant Cost in the Reference Plant.

The coal feed pump, likewise, reduces total plant cost by $ 65/kW not because of less expensive coal feed system to the gasifier, but because the coal flowrate decreases by 9 % resulting in reduced equipment sizes throughout the plant. The reduction in TPC on a $/kW basis (by only 3 %) is somewhat dampened, however, by decreased power output from the plant as a result of less coal feed.

Warm gas cleanup and ITM represent capital cost reductions from the cold gas cleanup and cryogenic ASU sections that they replace – reducing by $73/kW and $79/kW in section cost alone. However, that cost reduction is amplified throughout other plant sections by increased net power generated, reducing the entire plant cost by $23 1/kW and $124/kW, respectively.

Levelized Cost of Electricity

For each technology, process efficiency improvements (resulting in reduced fuel cost) and reduced total plant cost are reflected in the COE reductions illustrated in Figure 4-3. Capacity factor remains constant at 75 % in all reference plant cases.

Advanced syngas turbine technology has the most significant impact on COE reductions. Relative to the 7FA turbine, the advanced "F" frame, 2010-AST, and 2015-AST turbines reduce COE in the reference process by 11.3, 14.5, and 15.9 mills/kW-hr, respectively.

Warm gas cleanup, because it has lower capital cost, greater plant power output, and higher process efficiency than cold gas cleanup, represents about a 6.5 mills/kW-hr reduction in COE for partial warm gas cleanup, and about 8.1 mills/kW-hr reduction for full warm gas cleanup.

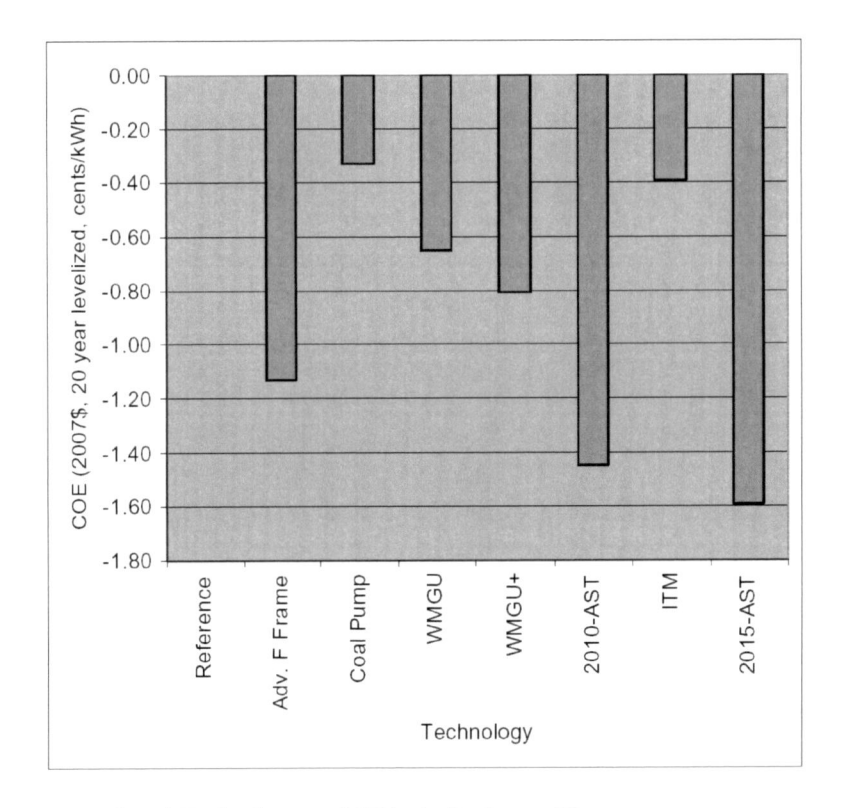

Figure 4-3. Impact of Each Technology on COE in the Reference Plant.

The $ 124/kW reduction in total plant cost of the ITM plays a greater role in COE reduction than does the 0.3 percentage point improvement in process efficiency; these combine for a 3.9 mills/kW-hr COE reduction from ITM technology. Once again, this reduction is relative to the reference plant with advanced "F" frame gas turbine (Case 2a).

The coal feed pump contributes about a 3.3 mills/kW-hr reduction in COE resulting from $65/kW reduction in total plant cost (despite reduced net power output) and a 1.9 percentage point process efficiency improvement that helps to reduce fuel cost.

4.2. Cumulative Impact of Advanced Technologies

Process Efficiency

As each technology is introduced to the composite process, the following graph shows the cumulative improvement in process performance. Cases that feature improved capacity factor do not contribute to performance efficiency because the capacity factor merely increases the percentage of on-stream operation.

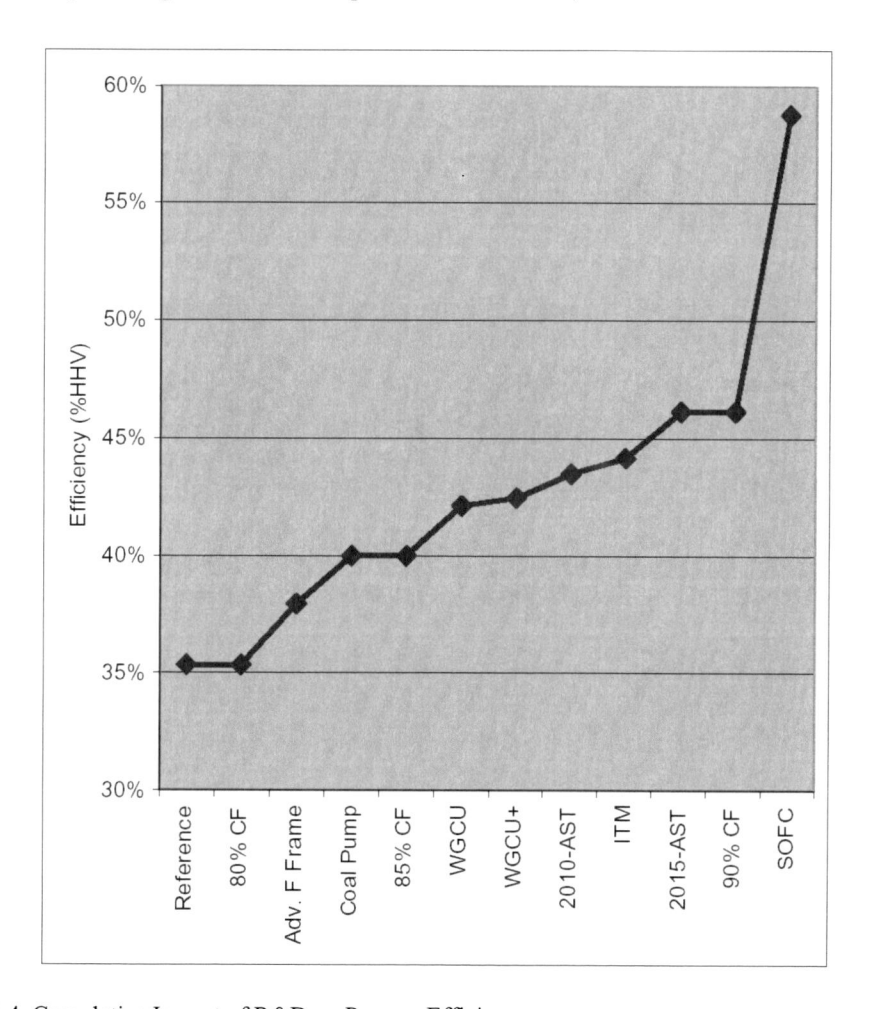

Figure 4-4. Cumulative Impact of R&D on Process Efficiency.

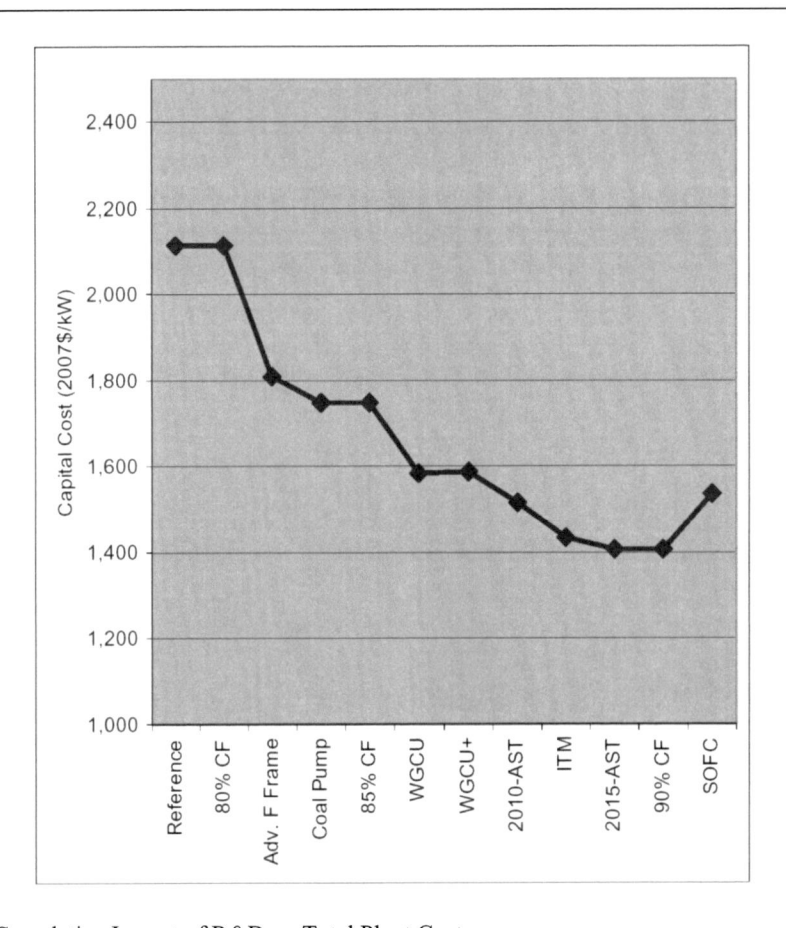

Figure 4-5. Cumulative Impact of R&D on Total Plant Cost.

The advanced "F" frame turbine and coal feed pump contribute 2.5 and 2.1 percentage point efficiency improvements, respectively. These are slightly greater than the sum of their individual efficiency improvements in the reference plant, so some synergy results from the combined technologies.

Partial warm gas cleanup also improves performance of the cumulative process (by 2.1 percentage points) more than it does the performance of the reference plant (2.0 percentage points). Full warm gas cleanup does not add very much more to performance in the cumulative process (only 0.3 percentage points) because elimination of the ammonia quench, which avoids condensing moisture from fuel gas in the slurry feed gasifier, does not represent as much of an advantage in a dry feed gasifier whose syngas has virtually no moisture in it.

The 2010-AST turbine, ITM, and 2015-AST each improve process efficiency (1.0, 0.7, and 2.0 percentage points, respectively) by slightly greater amount than they improved the reference plant (0.9, 0.3, and 1.7 percentage points, respectively). This again demonstrates some synergy resulting from combined technologies.

The integrated gasification solid oxide fuel cell process yields 58.8 % plant efficiency. This plant relies on a catalytic gasifier with very high (92.0 %) cold gas efficiency and full warm gas cleanup in order to avoid condensing moisture from syngas. Compared to the reference proces s, this represents a substantial 23.4 percentage point improvement in process efficiency. The high process efficiency is environmentally attractive because it reduces the production of CO_2 per megawatt of power produced. In addition, the sequestration-ready CO_2 stream that is produced holds promise for a superior process from the perspective of cost of CO_2 avoided for carbon capture scenarios.

Total Plant Cost

As each advanced technology is introduced to the composite process, total plant cost generally decreases as shown in Figure 4-5. Improved capacity factor has no effect on TPC, just as it had no effect on process efficiency.

The advanced "F" frame turbine has greatest effect of any technology on the cumulative TPC reduction ($304/kW); this is because of the large increase (150 MW) in net power output relative to the 7FA syngas turbine that is replaced. The incremental reduction from the 2010-AST turbine is $72/kW – not as dramatic a decrease because the power output with the 2010-AST is only 50 MW more than with the advanced "F" frame turbine. The incremental capital cost reduction from the 2015-AST turbine is only $27/kW; this is because there is a large decrease (197 MW) in net power output from the plant because the number of trains has been cut from two to one in order to maintain nominal plant output of 600 MW.

As in the reference plant, the warm gas cleanup and ITM have a lower capital cost than the technologies that they replace, but TPC on a $/kW basis further decreases because net power produced by the plant increases by about 50 MW as a result of each of these technologies. These technologies have an incremental cost reduction of $164/kW and $82/kW, respectively. The cost difference between partial warm gas cleanup and full warm gas cleanup is negligible.

The gasifier resulting from the coal feed pump (with dry feed) is only slightly less costly than the slurry feed gasifier (by $24 MM), but the process as a whole reduces in cost (by $80 MM) due to decreased coal flowrate which results in smaller equipment sizes throughout the plant. Considering the reduced power output of the coal feed pump plant by 23 MW, TPC reduces by $60/kW.

No systems analysis attempt was made to investigate an optimum solid oxide fuel cell process configuration; there is potential for further cost reduction resulting from possibly using ITM air separation, water gas shift of the fuel gas before it enters the fuel cell, an alternate gasifier such as Great Point Energy's bluegasTM that produces an all-methane syngas, or other similar process modifications that would likely decrease total plant cost. The increase by $ 129/kW over the Case 10 (90 % CF) advanced IGCC process is an artifact of the assumed capital costs of the fuel cell system and catalytic gasifier, and has considerable uncertainty at this time.

Cost of Electricity

As each new advanced technology is step-wise implemented in the advanced power system, the reduction in COE is represented in Figure 4-6. Effects of improved capacity factor are as significant as the other technology improvements that have yielded increased process efficiency and decreased capital cost. The increase to 80 % capacity factor results in a 4.0 mills/kW-hr decrease in COE, the increase to 85 % capacity factor results in a 2.9 mills/kW-hr decrease, and the increase to 90 % capacity factor results in a 2.0 mills/kW-hr decrease.

The advanced "F" frame syngas turbine provides the single greatest decrease in COE (10.7 mills/kW-hr) due to the 150 MW increase in net power output and 2.5 percentage point increase in plant efficiency due to air integration, improved turbine efficiency resulting from increased firing temperature and pressure ratio, and increased HRSG inlet temperature (allowing increased steam superheat and reheat temperatures).

Partial warm gas cleanup results in a 4.5 mills/kW-hr decrease in COE. Because of very low moisture content in the fuel gas, the novel ammonia and mercury removal units in the full warm gas cleanup case result in a very small improvement in process efficiency (leading to almost no change in fuel cost). There is no significant difference in TPC between partial and full warm gas cleanup, so as a result the COE changes very little between these cases.

The 2010-AST syngas turbine increases plant power output by 50 MW over that of the advanced "F" turbine, and therefore results in a $72/kW reduction in TPC. There is also a 1.0 percentage point improvement in process efficiency over the advanced "F", resulting in reduced fuel cost. Overall, the 2010-AST turbine decreases COE by 2.7 mills/kW-hr in the cumulative technologies plant.

The ITM increases plant output by 11 MW with a corresponding decrease in TPC by $43 MM, resulting in a $ 82/kW decrease in total plant cost. Although the efficiency improvement is only 0.7 percentage points, the decreased TPC translates to a 2.6 mills/kW-hr decrease in COE.

The 2015-AST syngas turbine has a much higher power rating than the 2010-AST, but the reduction from two trains to a single train decreases the net plant power output by 197 MW resulting in only a $27/kW reduction in TPC and, therefore, a 0.7 mills/kW-hr reduction in COE.

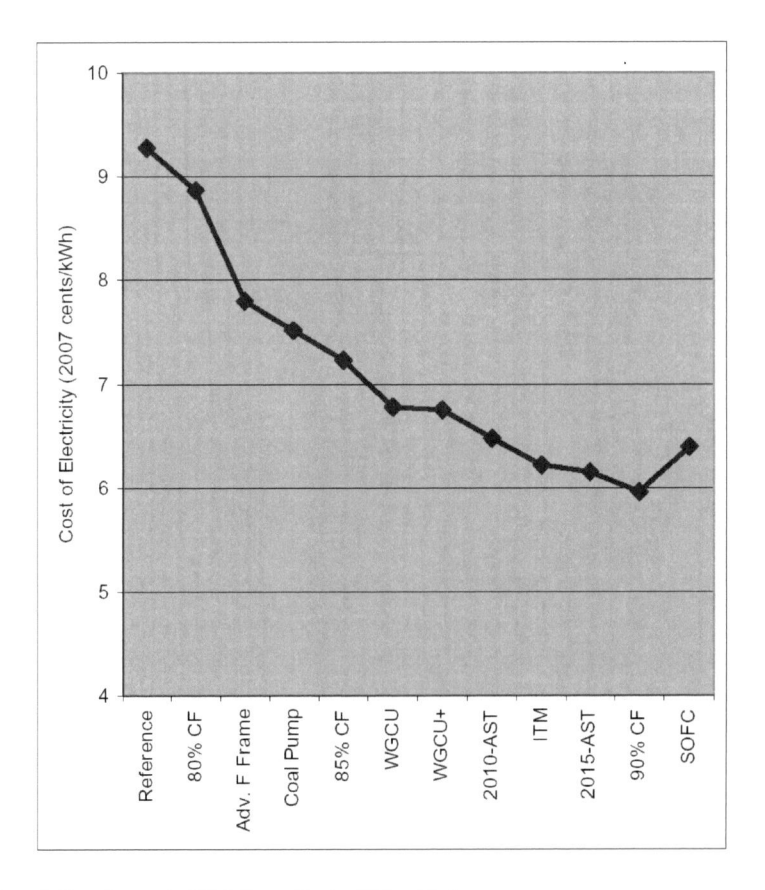

Figure 4-6. Cumulative Impact of R&D on Cost of Electricity.

The tremendous process efficiency (58.8 %) of the SOFC process makes its COE competitive with the most advanced Case 10 IGCC process, even before any systems analysis attempts at improved SOFC process configurations are made. The COE of 6.4 mills/kW-hr is based on an assumed fuel cell system TPC of $550/kW. With no further systems analysis attempt to improve process efficiency or power output, a TPC of $3 50/kW would be sufficient to reduce the SOFC cost of electricity to below that of the previous case with 2015-AST IGCC process and 90 % capacity factor. Although the particular SOFC configuration of Case 11 does not have the least COE, it has great potential for carbon capture scenarios because the CO_2 product stream is sequestration-ready.

5. SUMMARY

This pathway study evaluated anticipated process performance improvements and capital cost reductions resulting from advanced technology development sponsored by DOE. The study is presently confined to bituminous coal feedstock for process configurations that do not capture CO_2.

Advanced technology offers significant improvements in process efficiency. In the IGCC process alone, there is the potential for 11 percentage point improvement over the reference process. With SOFC technology, process improvements upwards of 24 percentage points are potentially achievable.

Capital cost reductions result not only from less expensive technology alternatives such as warm gas cleanup and ITM air separation, but also from increased power generation brought about by the advanced technology such as syngas turbines – resulting in cumulative total plant cost reductions by as much as $700/kW after all advanced technologies are implemented.

Improvements in process efficiency, reductions in capital and operating expense, and increase in capacity factor all contribute to decreased COE, projecting an overall decrease by more than 3 cents/kW-hr – or a decrease of 35 % in COE.

The advanced power systems technology pathway evaluated in this analysis covers a time span of about eighteen (18) years of technology development. Results of the analysis clearly indicate the importance of continued R&D, large scale testing, and integrated deployment so that future coal-based power plants will be capable of generating clean power with greater reliability and at significantly lower cost.

Aside from improved process efficiencies and reduced costs of electricity for non-capture power generation, these advanced technologies enable (1) production of high-value products such as hydrogen; (2) integration with solid oxide fuel cells, and (3) pre-combustion carbon capture at potentially lower cost than post-combustion alternatives. Volume 2 of this study will investigate applications of these DOE-sponsored advanced technologies in several different carbon capture configurations.

APPENDIX A. MODEL VALIDATIONS

A series of model validations, for both process performance and cost, is provided below to demonstrate that Noblis results presented in this report are consistent with those of other researchers and technology developers.

A.1. Case 2: Comparison with NETL Baseline Study

In order for the IGCC cases to be consistent with an established basis, Case 2 was given the same process configuration and design basis as NETL's Baseline Study Case 1. In the time since NETL's baseline case was developed, the radiant cooler temperature was determined to increase from 1,100 °F to 1,250 °F and the need for greater sour water stripper purge was identified in order to limit chloride concentration to 1,000 ppm. However, for the purpose of comparing Noblis' Aspen Plus process model with the baseline case, a simulation was developed using all the same process operating conditions as the Baseline Study case – identified as Case 2* in the Volume 1 Supplement to this report. The following table compares process performance.

Model results agree very closely with the Baseline Study. Note that the process efficiency of Case 2* is 38.3 % rather than 37.9 % in Case 2; this is due to the 1,100°F radiant cooler temperature (absorbing more heat from the raw syngas stream for steam generation) and lower sour water stripper heat duty (decreasing demand on the steam cycle).

Table A-2 compares capital and O&M costs with the Baseline Study Case 1. Although the TPC of the two cases agrees to within 0.2 %, the TPC on a $/kW basis is different by 0.8 %; this is due to slightly different net power production between the two cases (644 MW Noblis vs. 640 MW Baseline Study). This same difference in net power production is responsible for the slight difference in fuel cost and levelized cost of electricity. Notwithstanding, the results compare closely, and can be considered to agree.

A.2. Case 6: Comparison with Nexant

Model Validation

Nexant [6] reported a 3.6 percentage point process efficiency improvement from replacing cold gas cleanup with warm gas cleanup in an IGCC process with a slurry feed gasifier; this was somewhat greater than the 2.8 percentage point improvement between Noblis' Cases 0 and 6a. In a separate analysis [7], Noblis determined that the difference in process efficiency was due to a series of design features. The two most significant of these were:

- The quantity of reducing gas sent to the SCOT process for tail gas treatment and the ultimate disposition of the tail gas (whether discarded or recycled to Selexol).
- Utilization of low-quality heat from the low-temperature syngas cooling section – i.e. the assumed temperature cut-off at which heat was considered to be unrecoverable for purposes of steam generation.

Table A-1. Case 2 Process Performance Closely Agrees with NETL Baseline Case

	Baseline Study	Case 2*
Gas Turbine Power (MWe)	464	464
Fuel Gas Expander (MWe)	7	8
Steam Turbine Power (MWe)	299	301
Total Power Produced (MWe)	770	772
Auxiliary Power Use (MWe)	-130	-128
Net Power (MWe)	640	644
As-Received Coal Feed (lb/hr)	489,634	491,336
Net Heat Rate (Btu/kW-hr)	8,922	8,900
Net Plant Efficiency (HHV)	38.2 %	38.3 %

Table A-2. Case 2*: Capital and O&M Cost Comparison

	Baseline Study			Case 2*		
Capital Cost ($1,000)						
Plant Sections	**EPCC**	**TPC**	**TPC $/kW**	**EPCC**	**TPC**	**TPC $/kW**
1 Coal and Sorbent Handling	29,016	34,819	54	29,064	34,876	54
2 Coal and Sorbent Prep & Feed	45,072	55,887	87	45,148	56,024	87
3 Feedwater & Balance of Plant	30,687	37,580	59	30,630	37,506	58
4a Gasifier	213,000	273,078	426	213,027	273,112	424
4b Air Separation Unit	167,329	184,063	287	167,344	184,078	286
5a Gas Cleanup	108,066	129,980	203	107,726	129,573	201
5b CO2 Removal & Compression	0	0	0	0	0	0
6 Gas Turbine	103,787	119,642	187	103,491	119,302	185
7 HRSG	51,530	57,247	89	50,936	56,565	88
8 Steam Cycle and Turbines	59,162	67,201	105	59,002	67,038	104
9 Cooling Water System	23,258	28,032	44	22,793	27,245	42
10 Waste Solids Handling System	39,686	43,960	69	39,555	43,815	68
11 Accessory Electric Plant	58,625	69,826	109	58,591	69,781	108
12 Instrumentation & Control	19,154	23,382	37	19,032	23,240	36
13 Site Preparation	14,369	18,681	29	14,250	18,524	29
14 Buildings and Structures	15,080	17,541	27	15,078	17,540	27
Total	977,821	1,160,919	1,813	975,666	1,158,449	1,799
O&M Cost ($1,000)						
Fixed Costs	**Total**		**% EPCC**	**Total**		**% EPCC**
Labor	22,589		2.31	22,548		2.31

Table A-2. (Continued)

Variable Operating Costs*	Total	% EPCC	Total	% EPCC
Maintenance Materials	23,111	2.36	22,922	2.35
Water	1,767	0.18	1,710	0.18
Chemicals	1,339	0.14	1,304	0.13
Waste Disposal	2,919	0.30	2,918	0.30
Total Variable Costs	29,136	2.99	28,862	2.96
Total O&M Cost	51,725	5.29	51,410	5.27
Fuel Cost*	72,250	7.39	72,498	7.43
Discounted Cash Flow Results				
Total Plant Cost ($/kW)		1,813		1,799
Levelized Cost of Electricity ($/kW-hr)		0.0780		0.0774

*Includes 80% capacity factor

Other process differences and modeling assumptions included:

- Gasifier carbon conversion, pressure, and temperature
- Fuel gas heating value
- Gas turbine power and firing temperature
- Process heat losses
- Gas turbine air leakage and isentropic efficiencies
- CO_2 separation in the Selexol process (significant if the Claus tail gas is not recycled to Selexol)
- Auxiliary power requirements
- Fuel gas reheat temperature

Noblis' process parameters and modeling assumptions, to a great extent, are based on Case 1 from NETL's Baseline Study report. These design parameters are used throughout the pathway study for consistency – providing a common basis for comparison among advanced IGCC technologies. The remaining 0.8 percentage point in performance improvement between Noblis ' and Nexant's results due to warm gas cleanup can be explained in terms of different process configurations and modeling assumptions; Noblis was able to reproduce Nexant's results when the same modeling assumptions were used.

Cost Validation

The improvement in capital cost and levelized COE due to warm gas cleanup is compared below with Nexant's study to measure the agreement between the two studies and possible reasons for difference s. As already noted in the analysis of process efficiency improvement due to warm gas cleanup, multiple differences between Nexant's and Noblis' process parameters and modeling assumptions contribute to different process performance improvements attributed to warm gas cleanup.

The gas turbine (Nexant uses an advanced "F" turbine with no air integration whereas Noblis uses a 7FA turbine with no air integration) determines net power production. As observed from Noblis' results between the 7FA turbine in Case 0 and the advanced "F" frame turbine in Case 2a (both cases at 75 % capacity factor), total plant cost decreases by about

$304/kW and cost of electricity decreases by $0.011/kW-hr when using the larger advanced "F" frame syngas turbine.

Notwithstanding the differences in process parameters and gas turbine selection, some comparisons can be made between the Nexant and Noblis results to assess the cost benefit due to warm gas cleanup.

Both Nexant and Noblis base their cost estimates on January 2007 dollars. As seen in Table A- 3 below, the advanced "F" frame turbine produces more net power; when cold gas cleanup is replaced with warm gas cleanup, the power production incre ases by about 10 percent.

Comparing gas cleanup section costs, the Noblis costs on a $/kW basis are less than Nexant. This result does not correlate with total plant cost, which increases on a $/kW basis for the smaller 7FA plant; the gas cleanup section represents a greater percentage of TPC in Nexant's estimate than it does in Noblis' estimate. Nexant's warm gas cleanup cost decreases by $53/kW, or 17 % decrease over the cold gas cleanup cost. By comparison, Noblis' warm gas cleanup cost decreases by $73/kW, or 32% decrease over the cold gas cleanup cost.

The gas cleanup section affects other areas of the plant (such as equipment size scaling, steam cycle capacity, and balance of plant), so the comparison of total plant cost is also impacted by the transition from cold gas cleanup to warm gas cleanup. Nexant's TPC decreases by $269/kW, or a reduction by 14 % due to warm gas cleanup. Noblis predicts a $231/kW decrease in TPC, which corresponds to an 11 % reduction due to warm gas cleanup. These results, on a relative basis, agree better than the comparison on the basis of gas cleanup section alone.

Although there are discrepancies in TPC between the Nexant and Noblis cost estimates, both agree that a reduction of between 11-14 % in capital cost can be expected resulting from warm gas cleanup.

Nexant reported a reduction in COE from $0.0716/kW-hr to $0.0647/kW-hr resulting from the change from cold gas cleanup to warm gas cleanup; a reduction by 9.6 %. This was based on an 85 % capacity factor and fuel cost of $2.00/MMBtu.

Noblis' COE (based on 75 % capacity factor and fuel cost of $1.80/MMBtu) reduced from $0.0927/kW-hr to $0.0846/kW-hr. By changing the capacity factor and fuel cost to the same basis as Nexant, Noblis' results become $0.0885/kW-hr for cold gas cleanup reducing to $0.0801/kW-hr for warm gas cleanup – a savings of 9.1 % attributable to warm gas cleanup.

Table A-3. Comparisons of Improvements Due to Warm Gas Cleanup

	Nexant Case 3	Nexant Case 4	Noblis Case 0	Noblis Case 6a
	Adv. "F", Cold Gas Cleanup	Adv. "F", Warm Gas Cleanup	7FA, Cold Gas Cleanup	7FA, Warm Gas Cleanup
Net Power Production (MW)	585	641	487	541
Gas Cleanup Section ($/kW)	316	263	231	158
Total Plant Cost ($/kW)	1,904	1,635	2,113	1,882

The fact that Noblis' COE's are respectively greater than Nexant's – $0.0885/kW-hr vs. $0.0716/kW-hr for cold gas cleanup and $0.0801/kW-hr vs. $0.0647/kW-hr for warm gas cleanup – is expected because of the higher capital cost ($/kW) for the lower capacity 7FA plant compared to the advanced "F" turbine plant.

To summarize, given the differences in plant size, process performance improvement, uncertainties in cost estimation, and differences in economic assumptions, Nexant and Noblis independently estimate a reduction of between 11-14 % in capital cost, and a reduction of between 8-10 % in COE resulting from warm gas cleanup.

A.3. Case 11: Comparison with NETL

Model Validation

The fuel cell process configuration of Case 11 was developed by SAIC, and is described in NETL's report titled "The Benefits of SOFC for Coal-Based Power Generation" [11]. Becau se that process is fueled by Pittsburgh #8 coal, Noblis developed an Aspen Plus simulation bas ed on Pittsburgh #8 coal and then switched the feedstock to the s ame Illinois #6 coal used in all the previous pathway study cases and increased the net power production to the nominal 600 MW plant size. Table A-4 compares overall process performance to a case provided by SAIC[4] that is similar to Case 4 published in the NETL report.

Compared to the SAIC case, fuel cell power in Noblis' Pittsburgh #8 simulation decreases because of reduced fuel utilization (85 % vs. 89 %). This also reduces the work recovered by the cathode air expander (and cathode air compressor) due to reduced air flow through the fuel cell. The net power generated is very nearly the same, as is the net plant efficiency.

All power accounts increase in the Noblis Illinois #6 case because of the nominal 600 MWe net power production. Comparing the two Noblis cases using different coals, net plant efficiency decreases because of the change in fuel quality; gasifier cold gas efficiency decreases when going to the Illinois #6 coal due to increased fuel moisture content and decreased coal heating value.

Table A-4. Comparisons of Fuel Cell Process Using Different Coals

	SAIC Pitt#8	Noblis Pitt#8	Noblis Ill#6
Fuel Cell Power (MW)	439	431	517
Syngas Expander (MW)	18	18	22
Cathode Air Expander (MW)	197	157	208
Anode Exhaust Expander (MW)	101	108	132
Total Power Produced (MW)	755	714	879
Auxiliary Power Use (MW)	-248	-213	-276
Net Power (MW)	507	501	603
As-Received Coal Feed (lb/hr)	228,420	228,420	300,000
Net Heat Rate (Btu/kW-hr)	5,661	5,732	5,805
Net Plant Efficiency (HHV)	60.3 %	59.5 %	58.8 %
Gasifier Cold Gas Efficiency	91.1 %	93.1 %	92.0 %

Table A-5. Case 11: Capital and O&M Cost Validation

	NETL Case 4			Case 11		
Capital Cost ($1,000)						
Plant Sections	**EPCC**	**TPC**	**TPC $/kW**	**EPCC**	**TPC**	**TPC $/kW**
1 Coal and Catalyst Handling			93[6]	25,678	30,814	51
2 Coal and Catalyst Prep & Feed				33,550	41,428	69
3 Feedwater & Balance of Plant				17,805	21,649	36
4a Gasifier			311	123,853	155,335	258
4b Air Separation Unit			120	73,915	81,306	135
5a Gas Cleanup			134	55,056	66,351	110
5b CO2 Removal & Compression			0	0	0	0
6 Gas Turbine			0	0	0	0
7 Fuel Cell			392	278,195	333,836	554
8 Steam Cycle and Turbines			16	0	0	0
9 Cooling Water System				11,507	13,935	23
10 Waste Solids Handling System				32,217	35,692	59
11 Accessory Electric Plant				73,623	87,996	146
12 Instrumentation & Control				23,674	28,907	48
13 Site Preparation				14,480	18,823	31
14 Buildings and Structures			375[7]	8,594	10,097	17
Total			1,443	772,147	926,169	1,536
O&M Cost ($1,000)						
Fixed Costs		**Total**	**$/kW-hr**	**Total**		**$/kW-hr**
Labor			0.0068	18,039		0.0043
Variable Operating Costs*		**Total**	**$/kW-hr**	**Total**		**$/kW-hr**
Maintenance Materials				25,170		
Water				141		
Chemicals				3,410		
Fuel Cell Stack Replacement				15,853		0.0038
Waste Disposal				2,130		
Total Variable Costs			0.0056	46,704		0.0111
Total O&M Cost			0.0124	64,743		0.0153
Fuel Cost*			0.0099	44,266		0.0105
Discounted Cash Flow Results						
Total Plant Cost ($/kW)			1,443			1,536
Levelized Cost of Electricity ($/kW-hr)			0.062			0.0687

*Includes 80% capacity factor

Cost Validation

Table A-5 compares capital and operating costs with Case 4 from NETL's "The Benefits of SOFC for Coal-Based Power Generation". The fuel cell process of NETL Case 4 is very similar to the SAIC fuel cell process against which process performance was compared, so it is reasonable to compare relative costs with Noblis' Case 11.

NETL's Case 4 produces a net 523 MW of power using Pittsburgh #8 coal. It has an overall process efficiency of 62.0 %, with a small power contribution from a steam cycle (that was not part of SAIC's process that was used as the basis for this case). The ASU in NETL's Case 4 produces 95 % pure oxygen.

Noblis' capital cost estimates for coal and catalyst handling, preparation, and feed are greater than NETL's; this can be attributed to the uncertainty of the additional cost required for gasifier catalyst handling and feed systems. The same consideration also applies to the gasifier cost which includes the char/catalyst separation and coal/catalyst treatment systems; the Noblis gasifier cost estimate is 17 % less than NETL's estimate.

Noblis' ASU cost is 13 % higher than NETL's. The fact that Noblis' value is greater than NETL's is justifiable, considering that the oxygen purity is 99.5 % vs. 95 % in NETL's case.

Noblis' assumed TPC of $550/kW for the fuel cell system is 40 % greater than the $392/kW used in NETL's case. This more than accounts for the $93/kW difference in Total Plant Cost between the two estimates.

Noblis' fixed O&M cost is somewhat lower than NETL's; Noblis' estimate of 12 operators and technicians is based on a correlation used in the IGCC cases in which labor cost is a function of EPCC. NETL includes fuel cell stack replacement cost in fixed O&M. The larger net power production in Noblis' case is another reason for further decreasing fixed labor cost on a $/kW-hr basis; net power production in NETL's case is only 522 MW vs. 603 MW in Noblis' case.

For comparison purposes, the values listed in Table A-5are ba sed on 80 % capacity factor. Noblis' variable operating costs are significantly greater than NETL's. The two most significant cost accounts are maintenance materials and fuel cell stack replacement. Noblis assumed no change in the maintenance cost algorithm used in IGCC cases. Fuel cell stack replacement is assumed to cost $175/kW, with service life of 40,000 hours.[5] Noblis' chemical cost consists primarily of trona and ZnO sorbent costs, and does not include an estimate for gasifier catalyst cost. If Noblis' $0.0038/kW-hr (un-levelized) variable O&M cost for fuel cell stack replacement were moved to fixed O&M as in NETL's calculation, both Noblis' fixed O&M and variable O&M costs would be greater than NETL's. As shown under total O&M cost, Noblis' O&M cost is 23 % greater than NETL's. Noblis' resulting COE is 10 % greater than NETL's.

LIST OF REFERENCES

[1]　Cost and Performance Baseline for Fossil Energy Plants. Volume 1: Bituminous Coal and Natural Gas to Electricity. (May, 2007). Department of Energy, National Energy Technology Laboratory. DOE/NETL-2007/1281.

[2]　"IGCC: What's GE Up To?" (October 13, 2005). Norm Shilling, General Electric. American Coal Council 2005 Coal Market Strategies.

[3]　Quality Guidelines for Energy System Studies. Work in Progress. U.S. Department of Energy, National Energy Technology Laboratory.

[4]　Power Systems Financial Model Version 5.0 Users' Guide. (September, 2006). Nexant. Prepared for U.S. Department of Energy, National Energy Technology Laboratory.

[5] "Capital Charge Factors and Levelization Factors – Tables for QGESS" provided in E-mail from John Wimer, NETL. (June 5, 2007).

[6] "Preliminary Feasibility Analysis of RTI Warm Gas Cleanup (WGCC) Technology." (June 2007). Prepared by Nexant for RTI International.

[7] "Reconciliation of Noblis Warm Gas Cleanup Results with Nexant's Results." (January 2008). Technical note prepared by Noblis for NETL.

[8] "ITM Oxygen Offers 15 % More Power and Over 10 % Better Plant Efficiency." Harry Jaeger. *Gas Turbine World*. Volume *38*, Number 1.

[9] "Advanced Technology Options for FutureGen". (March 30, 2006). Gary J. Stiegel, National Energy Technology Laboratory. Paper presented at Pittsburgh Coal Conference. Pittsburgh, PA.

[10] "New Gas Turbine Integration Options for ITM Oxygen in Gasification Applications". (October 17,2007). Stein, V., Armstrong, P. and T. Foster. Air Products and Chemicals, Inc. Gasification Technologies 2007. San Francisco, CA.

[11] "The Benefits of SOFC for Coal-Based Power Generation." (October 30, 2007). Report Prepared by E. Grol, J. DiPietro, and J. Thijssen for Wayne Surdoval. National Energy Technology Laboratory.

[12] "Successful Continuous Injection of Coal into Gasification and PFBC System Operating Pressures Exceeding 500 psi – DOE Funded Program Results". Saunders, T., Aldred, D., and Rutkowski, M. Gasification Technology Council Conference. San Francisco, CA. (October 9- 12, 2005).

[13] "Analysis of Stamet Pump for IGCC Applications". (December 2005). Topical Report. DOE/NETL-40 1/121905.

[14] "Compact Gasifier 90% Smaller and Half the Cost of Conventional Units." (2008). Gas Turbine World. Volume 38, No. 1.

[15] "Successful Continuous Injection of Coal into Gasification and PFBC System Operating Pressures Exceeding 500 psi – DOE Funded Program Results." T. Saunders (Stamet), D. Aldred (Stamet), and M. Rutkowski (Parsons). Gasification Technology Council Conference, San Francisco.

[16] "Design/Economic Analysis of RTI Warm Gas Cleanup Technology." (February 12, 2007). Presentation by Nexant at Project Review Meeting.

[17] "Novel Integrated Oxygen Supply Technology for Gasification." (November, 2006). Armstrong, P.A., Repasky, J.M., Rollins, W.S., and Stein, E.E. Paper presented at POWER-GEN International 2006. Orlando, Florida.

LIST OF ACRONYMS

AST	Advanced Syngas Turbine
ASU	Air Separation Unit
BEC	Bare Erected Cost
BOP	Balance of Plant
CCF	Capital Charge Factor
CF	Capacity Factor
COE	Cost of Electricity

COS	Carbonyl Sulfide
DB	Double-Declining Balance
DCF	Discounted Cash Flow
DOE	Department of Energy
DSRP	Direct Sulfur Reduction Process
EPCC	Engineering, Procurement, and Construction Cost
FYC	First Year Operating Costs
HHV	Higher Heating Value
HP	High Pressure
HRSG	Heat Recovery Steam Generator
IGCC	Integrated Gasification Combined Cycle
IGFC	Integrated Gasification Fuel Cell
IOU	Investor-Owned Utility
IP	Intermediate Pressure
ITM	Ion Transport Membrane
kW	kilowatt
kW-hr	kilowatt-hour
LF	Levelization Factor
LHV	Lower Heating Value
MM	million
MW	megawatt
MWe	megawatt - electric
MWh	megawatt hour
NETL	National Energy Technology Laboratory
NOx	Nitrogen Oxides
O&M	Operating and Maintenance
PSFM	Power Systems Financial Model
QGESS	Quality Guidelines for Energy System Studies
R&D	Research and Development
RAM	Reliability, Availability, and Maintainability
SOFC	Solid Oxide Fuel Cell
SOx	Sulfur Oxides
sTPD	Standard Tons per Day
TDS	Transport Desulfurizer
TPC	Total Plant Cost
TRC	Total Required Capital
WGCU	Warm Gas Cleanup

End Notes

[1] As an alternative to the PSFM, a separate Excel spreadsheet model was developed for economic analysis in this study; it was validated against the PSFM to verify that the calculations were implemented correctly. The spreadsheet model was developed to contain both capital cost algorithms and DCF calculations in a single file.

[2] Warm gas cleanup chemical costs were verified by personal communication with Brian Turk, RTI.

[3] A goal of the fuel cell program is to develop a power system with cost equal to or less than an equivalent natural gas combined cycle power system. That cost, in January 2007 dollars, was estimated to be $ 550/kW.

[4] Personal correspondence with D. Keairns and R. Newby.

[5] Replacement cost and service life were provided by personal communication from Wayne Surdoval, NETL.

[6] Coal handling is presumed to also include the preparation and feed systems that are evaluated separately in Noblis' cost estimate.

[7] This number represents an account named "Other" in NETL's report. It is presumed to include feedwater and BOP, cooling water system, waste solids handling system, accessory electric plant, instrumentation & control, site preparation, and buildings and structures.

In: Advanced Power Systems using Bituminous Coal
Editor: Daniel A. Lakatos

ISBN: 978-1-61324-670-2
© 2011 Nova Science Publishers, Inc.

Chapter 2

A Pathway Study Focused on Carbon Capture Advanced Power Systems R&D Using Bituminous Coal - Volume 2[*]

United States Department of Energy

List of Acronyms and Abbreviations

AHT	Advanced Hydrogen Turbine
AST	Advanced Syngas Turbine
ASU	Air Separation Unit
BEC	Bare Erected Cost
CCF	Capital Charge Factor
CF	Capacity Factor
COE	Cost of Electricity
COS	Carbonyl Sulfide
DOE	Department of Energy
DSRP	Direct Sulfur Reduction Process
EPCC	Engineering, Procurement, and Construction Cost
FYC	First year variable operating costs
HHV	Higher Heating Value
HRSG	Heat Recovery Steam Generator
IGCC	Integrated Gasification Combined Cycle
IGFC	Integrated Gasification Fuel Cell
ITM	Ion Transport Membrane
kW	kilowatt
kW-hr	kilowatt-hour
LF	Levelization factor
LHV	Lower Heating Value

[*] This is an edited, reformatted and augmented version of United States Department of Energy publication DOE/NETL-2009/1389, dated November 2010.

MM	million
MW	megawatt
MWh	megawatt hour
NETL	National Energy Technology Laboratory
O&M	Operating and Maintenance
R&D	Research and Development
RAM	Reliability, Availability, and Maintainability
SOFC	Solid Oxide Fuel Cell
TASC	Total As-Spent Cost
TOC	Total Overnight Cost
TPC	Total Plant Cost
TRC	Total Required Capital
TS&M	Transportation, Storage, and Monitoring of CO_2
WGCU	Warm Gas Cleanup

EXECUTIVE SUMMARY

The United States Department of Energy's (DOE) Strategic Center for Coal funds research and development (R&D) with the objective to improve the efficiency and reduce the cost of advanced power systems. In order to evaluate the benefits of on-going R&D, Noblis utilized their energy systems analysis capabilities and Aspen Plus computer simulation models to quantify the impact of successful federally-funded R&D on future power systems configurations.

This report represents Volume 2 of a two-volume Pathway Study in which a variety of process configurations that produce electric power from bituminous coal are analyzed. While Volume 1 [1] focuses on non-carbon capture process scenarios, Volume 2 addresses pre-combustion carbon capture scenarios. Each volume begins with a reference integrated gasification combined cycle (IGCC) plant using conventional technology, and a series of process modifications are made to represent commercialization of advanced technologies. Impacts of each technology on both process performance and cost are evaluated. In this manner, DOE can measure and prioritize the contribution of its R&D program to future power systems technology.

Advanced technologies within DOE's R&D program include:

- Three models of advanced hydrogen turbines (AHT)
- Dry coal feed pump
- Improved capacity factor resulting from equipment design and operating experience Warm gas cleanup (WGCU)
- Hydrogen membrane
- Ion transport membrane (ITM) for oxygen production
- Pressurized solid oxide fuel cell (SOFC)

Compared to non-capture technology, requirements for carbon capture impose both performance and cost penalties. The penalties are primarily the result of the parasitic energy and the capital cost of additional technology needed to separate CO_2 from process streams and compress the CO_2 to a pressure suitable for pipeline transport to a sequestration site. Advanced technology not only improves process performance and reduces the cost of electricity, but it also helps to reduce the incremental cost of carbon capture. Assuming R& D success in terms of performance and cost, the conceptual process configurations for each of these advanced technologies follow a pathway to an advanced IGCC plant with 90 % carbon capture that (1) is 9.6 percentage points greater in efficiency, and (2) reduces the 20-yr levelized cost of electricity (COE) by greater than 35 % relative to the reference carbon capture IGCC plant. An alternate pathway provided by an advanced integrated gasification fuel cell (IGFC) plant provides a high efficiency, near-100 % capture solution at a COE similar to that of the advanced IGCC.

Reference Plant Design Basis

The reference non-capture IGCC configuration from Volume 1 uses conventional technology from the year 2003 that features a single-stage slurry feed gasifier with radiant-only gas cooler followed by Selexol acid gas removal, a 7FA syngas turbine, and conventional three-pressure level steam cycle. Gasifier oxygen is provided by a cryogenic air separation unit (ASU). Process operation assumes a 75 % capacity factor.

In this Volume 2, to obtain the reference IGCC configuration with carbon capture the non- capture configuration is modified by: (1) converting sour syngas to hydrogen-rich fuel through water gas shift; (2) changing the acid gas removal section to conventional two-stage Selexol to accomplish CO_2 separation; (3) adding a CO_2 compression section, and (4) modifying the 7FA-based turbine to be powered by the hydrogen-rich fuel. The capacity factor is increased to 80 % to represent operating experience to date gained through DOE's Clean Coal Program as well as to account for improved reliability and availability expected to occur by the time that carbon capture cases are deployed. In the reference plant configuration, addition of carbon capture results in an efficiency reduction of 5 percentage points and a capital cost increase of $600/kW compared to its non-capture counterpart.

Process Improvements from Advanced Technologies

A series of conceptual process configurations with carbon capture that produce electric power from bituminous coal is analyzed to determine the potential performance improvements and cost reductions resulting from successful R&D of advanced technology. These process configurations are listed in Table ES-1. The white blocks represent existing, commercially available technologies while the colored blocks represent advanced emerging technologies. Each advanced technology is implemented and evaluated in a composite process in the order in which demonstration-readiness is anticipated. This allows assessment of the cumulative improvements in process performance and cost over time. The majority of the technologies are evaluated in the context of an IGCC plant. The single IGFC case represents an advanced process configuration that occurs later in the commercialization timeline, incorporating technologies that are of specific value to an IGFC plant.

Table ES-1. Carbon Capture Power System Technology Development

Case Title	Gas Turbine	Coal Feed System / Gasifier	Capacity Factor	Gas Clean Up	CO$_2$ Separation	Oxygen Production
Reference IGCC	7FA	Slurry Feed	80% CF	2-Stage Selexol		Cryogenic
Adv "F" Turbine	Adv "F"					Air
Coal Feed Pump		Coal				Separation
85% CF		Feed	85% CF			Unit
WGCU/Selexol		Pump		WGCU	Selexol	(ASU)
WGCU/ H$_2$Membrane					High	
AHT-1 Turbine	AHT-1				Temp	
ITM					Hydrogen	ITM
AHT-2 Turbine	AHT-2				Membrane	
90% CF			90% CF			
Advanced IGFC	Pressurized SOFC	Catalytic Gasifier	90% CF	WGCU	SOFC + Oxycombustion	Cryogenic ASU

Table ES-2. Cumulative Cost and Performance Impact of R&D for Gasification-Based Power Generation

Case Title	Efficiency (% HHV)	Delta* Efficiency (% points)	TPC** ($/kW)	Delta* TPC** ($/kW)	20-yr Levelized COE (¢/kW-hr)	Delta* COE (¢/kW-hr)
Reference IGCC	30.4	0	2,718	0	11.48	0
Adv "F" Turbine	31.7	1.3	2,472	-246	10.64	-0.84
Coal Feed Pump	32.5	0.8	2,465	-7	10.54	-0.10
85% CF	32.5	0.0	2,465	0	10.14	-0.40
WGCU/Selexol	33.3	0.8	2,425	-40	10.00	-0.14
WGCU/H$_2$Membrane	36.2	2.9	2,047	-378	8.80	-1.20
AHT-1 Turbine	38.0	1.8	1,855	-192	8.14	-0.66
ITM	38.3	0.3	1,724	-131	7.74	-0.40
AHT-2 Turbine	40.0	1.7	1,683	-41	7.61	-0.13
90% CF	40.0	0.0	1,683	0	7.36	-0.25
IGCC Pathway		+9.6%pts (+32%)		-1,035 (-38%)		-4.12 (-36%)
Advanced IGFC	56.3	+26%pts +85%	1,759	-959 (-35%)	7.45	-4.03 (-35%)

* Delta shown is the incremental change as each new technology is added to previous case configuration

** TPC is reported in January 2007 dollars and excludes owner's costs

Cost and Performance Impact of Advanced Technologies

See Appendix A for NETL 's update to capital costs and COE.[1]

Table ES-2 summarizes the results of the analysis as each new technology is added to the pathway, highlighting the increase in efficiency and decrease in total plant cost (TPC) and 20-year levelized COE. The delta for each metric provides an estimate of the incremental benefits of successful R&D for each technology. Turbine advancements contribute 50 % of the efficiency improvement and 40 % of the reduction in COE. The combined benefits of WGCU and the hydrogen membrane contribute 40 % of the efficiency benefit and 30 % of the COE reduction. The remaining benefits are due to a combination of the coal feed pump, ITM, and research efforts to improve plant availability. Details on the contributions of each advanced technology are provided in the following paragraphs.

Advanced Turbines

Advanced turbines contribute 4.8 (1.3+1.8+1.7) percentage points to increased process efficiency due to the combination of (1) improved engine performance at increasingly higher pressure ratios and firing temperatures, (2) air integration that reduces auxiliary load of the main air compressor, and (3) increased turbine exit temperature, which improves heat recovery from the heat recovery steam generator (HRSG).

Advanced hydrogen turbines also significantly reduce total plant cost. Although the cost of the turbine itself increases due to increased size, TPC on a $/kW basis decreases because of increased net plant power. The advanced "F" turbine and the first generation (AHT-1)[2] turbine contribute significant COE reductions – a total of 15 (8.4+6.6) mills/kW-hr. To maintain a nominal 600 MW plant size (the basis of this study), there is a reduction from two process trains to a single process train for the next generation (AHT-2) turbine. The reverse economy of scale associated with the train reduction translates into a minor decrease (1.3 mills/kW-hr) in COE.

If instead two trains are utilized, resulting in a 1 GW capacity unit, the COE change associated with incorporation of the advanced turbine is 8.2 mills/kW-hr (an 11 % reduction). Table ES-2 reports the costs corresponding to the more conservative single-train, nominal 600 MW configuration.

Coal Feed Pump

The coal feed pump increases the gasifier cold gas efficiency by eliminating the need to evaporate water in a slurry-fed gasifier. This benefit is somewhat countered by a higher steam requirement for the water gas shift reaction than was needed with a slurry feed. The resulting efficiency benefit is 0.8 percentage points.

The minor change in cost of equipment, coupled with a small reduction in net power associated with the coal feed pump, results in a negligible impact on TPC and COE.

Warm Gas Cleanup and Hydrogen Membrane

Warm gas cleanup (with Selexol CO_2 capture) improves process efficiency over cold gas cleanup in the carbon capture scenario as the result of eliminating the sour water stripper reboiler duty. However, coupling warm gas cleanup with the hydrogen membrane contributes even more increase in process efficiency by eliminating the Selexol regeneration steam requirements and auxiliary power, and also by producing CO_2 at elevated pressure – reducing CO_2 compressor load.

The cost of warm gas desulfurization is projected to be less than single-stage Selexol, which partly accounts for the decrease in TPC of the WGCU+Selexol configuration. An even greater reduction in TPC results with the addition of a hydrogen membrane that replaces the second- stage Selexol absorber for CO_2 capture. Furthermore, the cost of CO_2 compression is much less in the WGCU+Membrane case than any of the previous carbon capture cases due to the higher pressure at which CO_2 is produced from the H_2 membrane. Finally, when the added net power generation (made possible by eliminating the sour water stripper and Selexol reboilers and reducing CO_2 compression parasitic losses) is divided into the already-reduced TPC, the cost of the WGCU+Membrane case decreases by $41 8/kW (40+378) relative to the cold gas cleanup configuration. The COE benefit follows suit, decreasing by 13.4 mills/kW-hr (1.4+12.0).

Ion Transport Membrane

The ITM does not contribute strongly to process performance; its primary benefit is decreased capital cost of oxygen production. The ITM is predicted to reduce TPC by $131 /kW and the COE by 4.0 mills/kW-hr.

Reliability, Availability, and Maintainability (RAM)

Anticipated improvements in process RAM due to R&D in areas such as vessel refractories, improved sensors and advanced process controls are modeled as an increase in capacity factor. Although increased capacity factor does not influence either process efficiency or TPC, the added on-stream plant operation decreases COE by a total of 6.5 mills/kW-hr (4.0+2.5).

Pressurized Solid Oxide Fuel Cell

The pressurized solid oxide fuel cell case is capable of a process efficiency that approaches 60 %. The catalytic gasifier, with high methane content in the syngas, operates with a cold gas efficiency in excess of 90 %. Conversion of chemical energy within the fuel cell, as opposed to thermal and mechanical energy conversion in an IGCC process, enables the higher process efficiency obtained in the IGFC case.

Despite much higher process efficiency, higher capital costs of the IGFC process relative to IGCC result in a TPC and COE that are slightly greater than the most advanced IGCC configuration with carbon capture. However, the SOFC case results in nearly 100 % CO_2 removal compared to the 90 % capture of the IGCC.

Comparison to Non-Capture Scenarios

Figure ES-1 depicts the cumulative improvements in process efficiency, TPC, and COE as each technology is introduced for the carbon capture cases described in this study and the non-capture cases from Volume 1. The overall efficiency improvement for the IGCC non-capture pathway is 10.7 percentage points, slightly greater than the 9.6 percentage points achieved in the carbon capture cases. TPC (on a $/kW basis) and COE decrease by approximately 33 % in the non- capture IGCC cases, compared to 38 % and 36 % reduction in TPC and COE for the carbon capture cases, respectively.

The bottom of the shaded bars on the TPC and COE pathways illustrate the impact of the AHT-2 turbine if two turbine trains were built. That installation would exceed the nominal 600 MW plant size for this study, but the point serves to illustrate the effect of economy of scale on process economics.

While warm gas cleanup results in greater process efficiency improvement for the carbon capture scenario, its impact is especially pronounced in terms of TPC and COE. The cost differential between warm gas cleanup and cold gas cleanup is greater (resulting in more cost reduction) in the carbon capture scenario due to the additional Selexol absorber. In addition, the cost of CO_2 compression is much less in the WGCU+Membrane case than any of the previous carbon capture cases due to the higher pressure at which CO_2 is produced from the H_2 membrane. Finally, when the added net power generation (made possible by eliminating sour water stripper and Selexol reboiler duties and reduced CO_2 compression parasitic loss) is divided into the already-reduced TPC, the cost of the warm gas cleanup cases on a $/kW basis becomes $418/kW less than the cold gas cleanup carbon capture scenario, and COE decreases by more than 13 %. By comparison, warm gas cleanup in the non-capture scenario decreases TPC by $ 161/kW and COE by almost 7 %.

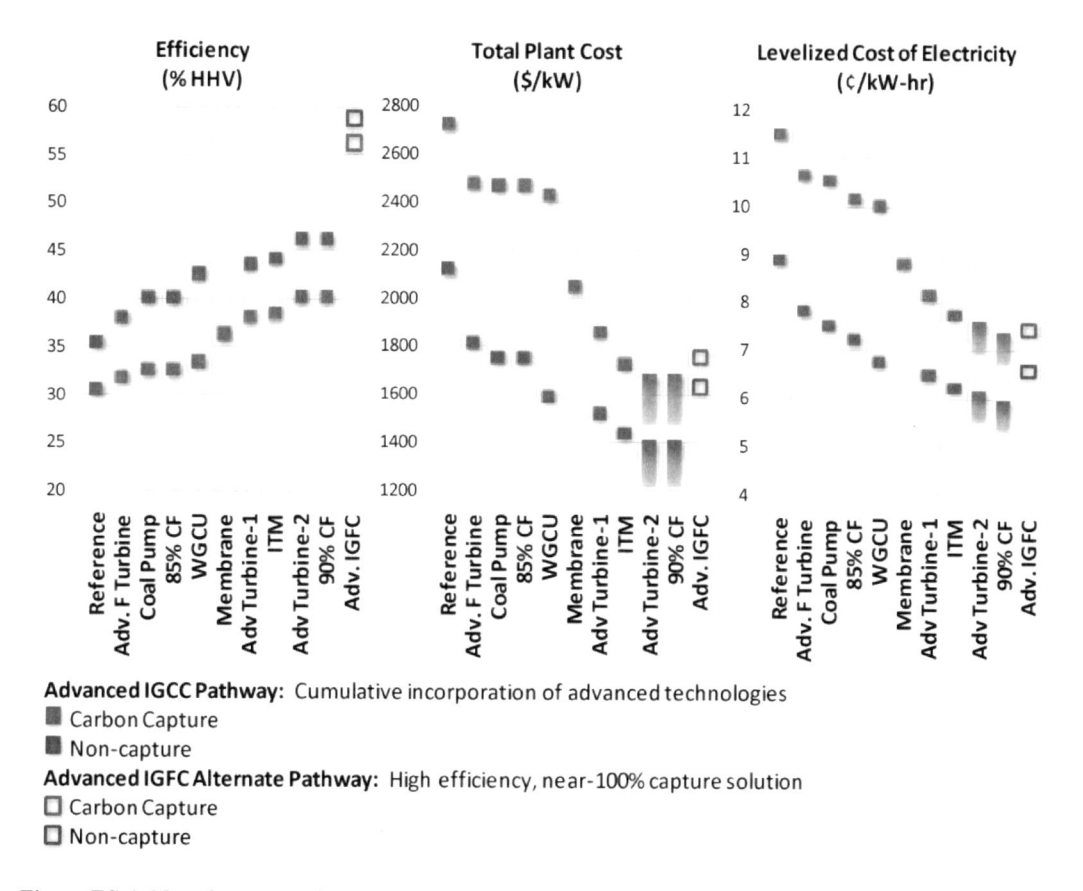

Figure ES-1. Non-Capture and Carbon Capture Pathway Results.

The coal feed pump makes a greater contribution to process efficiency and cost improvement in the non-capture scenario (2.1 percentage point efficiency increase and 4 % reduction in COE) than in the carbon capture scenario (0.8 percentage point efficiency increase and 1 % COE reduction). The coal feed pump increases process efficiency by eliminating the need to evaporate water in a slurry-fed gasifier. In the non-capture scenario with cold gas cleanup, that moisture is condensed and most of the latent heat is unrecoverable because of the low condensation temperature. In the carbon capture scenario with cold gas cleanup, on the other hand, moisture is needed for water gas shift; so whether the moisture is provided by slurry water or addition of shift steam (following a dry feed gasifier), the coal feed pump doesn't have as much of an impact on process efficiency.

The ITM is seen to reduce TPC by relatively more in the carbon capture scenario ($131/kW) than in the non-capture scenario ($82/kW). With an increase in coal feed rate to generate hydrogen turbine fuel compared to syngas turbine fuel, the significance of the air separation unit increases. This is because, with increased oxygen demand in the carbon capture cases, the capital cost savings represented by the less-expensive ITM compared to cryogenic ASU has a greater impact on reducing cost.

COE in the non-capture SOFC case increases by 11 % over that of the most advanced non- capture IGCC technology; this is due to a higher TPC that, even despite much higher process efficiency, results in a COE that is greater than IGCC by 6.6 mills/kW-hr. In the carbon capture scenario the sequestration-ready CO_2 stream from the SOFC incurs minimal incremental capital cost. The resulting COE, aided by 56.3 % process efficiency, is just 0.9 mills/kW-hr (1 %) greater than the most advanced carbon capture IGCC configuration.

DOE's Carbon Capture Targets

DOE's advanced power generation program goals are to achieve 90 % carbon capture while maintaining less than 10 % increase in COE over a 2003 reference IGCC plant having no carbon capture. That reference plant is represented in Case 0 in Volume 1 of this Pathway Study. At 75 % capacity factor the COE of that plant is 9.3 ¢/kW-hr, so DOE's cost target for carbon capture is 10 % greater, or 10.2 ¢/kW-hr.

From Figure ES-1 above, DOE's carbon capture target should be met early in the pathway, specifically by the case with 85 % capacity factor. Other process features of that case include advanced "F" hydrogen turbine, dry feed gasifier, cryogenic ASU, and cold gas cleanup.

All subsequent technology advancements will help to exceed DOE's program goals. By achieving the ultimate, most advanced IGCC and IGFC technologies projected in Figure ES-1, DOE could realize a 20 % *reduction* in COE over the 2003 reference IGCC plant having no carbon capture. The enabling technologies to achieve that improvement include:

- Advanced hydrogen turbines
- Coal feed pump
- Improved RAM
- Warm gas cleanup
- Hydrogen membrane

- ITM
- Pressurized SOFC with catalytic gasifier

The technology pathway evaluated in this study covers a time span of about 18 years of technology development. Results of the analysis clearly indicate the importance of continued R&D, large scale testing, and integrated deployment so that future coal-based power plants will be capable of generating clean power with greater reliability and at significantly lower cost.

Aside from improved process efficiencies and reduced costs of electricity for both non-capture and carbon capture power generation alike, these advanced technologies enable (1) production of high-value products such as hydrogen, (2) integration with solid oxide fuel cells, and (3) precombustion carbon capture projected at lower cost than post-combustion alternatives.

1. INTRODUCTION

The United States Department of Energy's (DOE) Strategic Center for Coal funds research and development (R&D) whose objective is to improve the efficiency and reduce the cost of advanced power systems. In order to evaluate the benefits of on-going R&D, Noblis utilized their energy systems analysis capabilities and Aspen Plus computer simulation models to quantify the impact of successful federally-funded R&D on future power systems configurations.

This report represents Volume 2 of a two-volume Pathway Study in which a variety of process configurations that produce electric power from bituminous coal are analyzed. While Volume 1 [1] focuses on non-carbon capture process scenarios, Volume 2 addresses pre-combustion carbon capture scenarios. Each analysis begins with a reference integrated gasification combined cycle (IGCC) plant using conventional technology, and a series of process modifications are made to represent commercialization of advanced technologies. Impacts of each technology on both process performance and cost are evaluated. In this manner, DOE can measure and prioritize the contribution of its R&D program to future power systems technology.

The advanced technologies that are examined in this volume include:

- Three models of advanced hydrogen turbines (AHT)
- Coal feed pump
- Improved capacity factor resulting from equipment design and operating experience Warm gas cleanup (WGCU)
- Hydrogen membrane for H_2 separation
- Ion transport membrane (ITM) for oxygen production
- Pressurized solid oxide fuel cell (SOFC) with catalytic gasifier

Compared to non-capture technology, requirements for carbon capture impose both performance and cost penalties. The penalties are primarily the result of the parasitic energy and the capital cost of additional technology needed to separate CO_2 from process streams and compress the CO_2 to a pressure suitable for pipeline transport to a sequestration site. Section 4 of this report compares the pathways of non-capture versus carbon capture power generation. As will be shown, advanced technology not only improves process performance and reduces the cost of electricity but it also helps to reduce the incremental cost of carbon capture.

2. PATHWAY STUDY BASIS

The design basis of NETL's Baseline Study [2] was adopted so that results from this pathway study would be consistent with established results. In general, all cases are based on a nominal plant size of 600 MW net power. A process flow diagram of the reference carbon capture case is provided in Figure 2-1. The process includes two 7FA hydrogen turbines and a steam cycle operating at 1,800 psig with 1,000 $^{\circ}$F steam superheat and 1,000 $^{\circ}$F steam reheat. The as- received Illinois #6 bituminous coal feed has a higher heating value of 13,126 Btu/lb (dry basis). Ultimate and proximate analyses of the coal are presented in Table 2-1.

Table 2-1. Bituminous Coal Analysis

Proximate Analysis As-Received (wt %)	
Moisture	11.12
Ash	9.70
Volatile Matter	34.99
Fixed Carbon	44.19
Ultimate Analysis Dry Basis (wt %)	
Ash	10.91
Carbon	71.72
Hydrogen	5.06
Nitrogen	1.41
Chlorine	0.33
Sulfur	2.82
Oxygen	7.75
Total	100.00
HHV (Btu/lb)	13,126

2.1. Process Description

A cryogenic air separation unit (ASU) provides oxygen for the single-stage, slurry feed, oxygen- blown gasifier. The ASU is sized to provide sufficient oxygen to the gasifier, plus a small slipstream of oxygen used in the Claus furnace for acid gas treatment. Most of the N_2 by-product can be compressed and injected into the topping combustor of the hydrogen turbine; the exact amount is determined by the turbine power rating, which is regulated to 192 MW per unit.

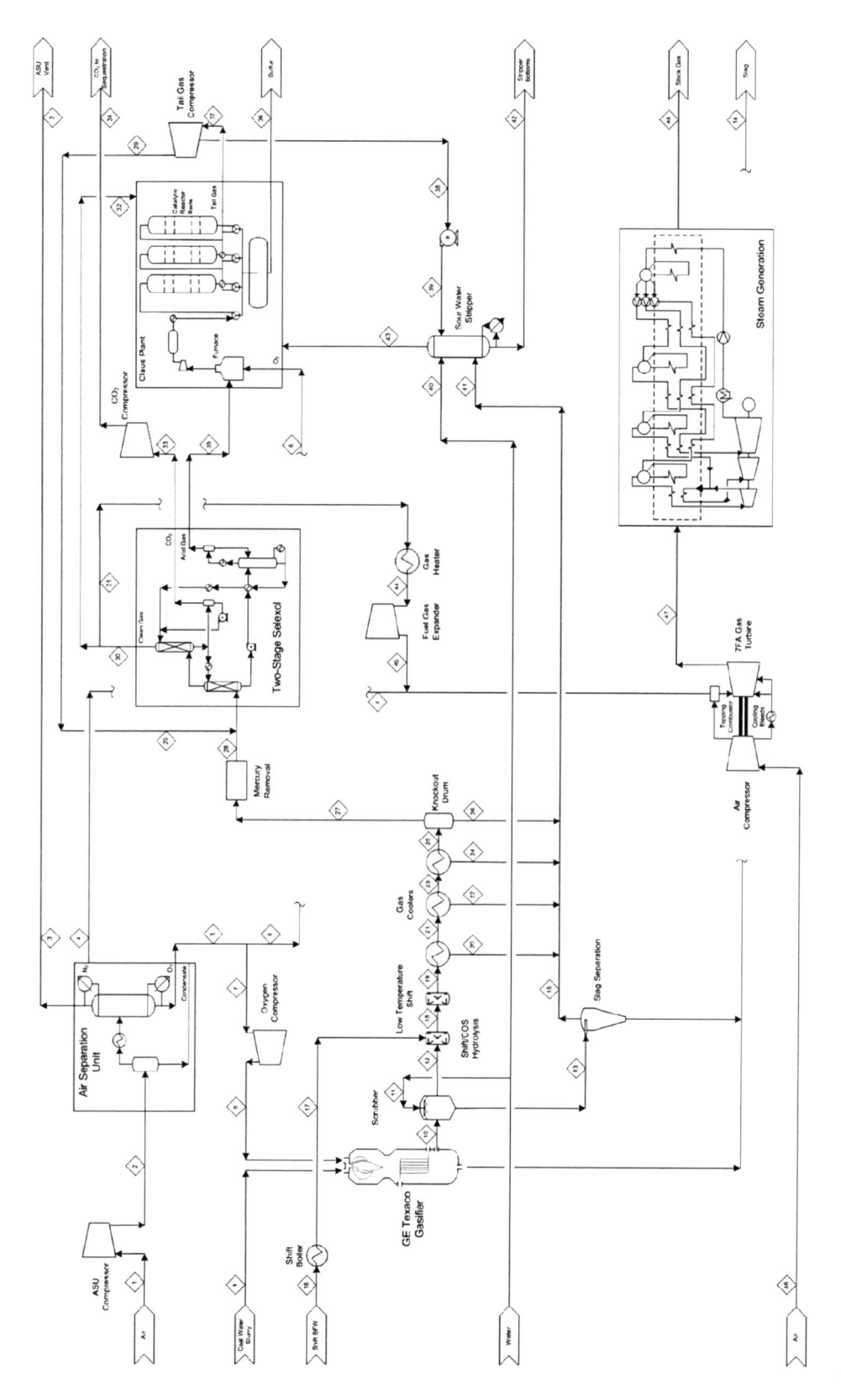

Figure 2-1. Process Flow Diagram of the Reference Carbon Capture Case.

Although the gasifier exceeds 2,400 °F during operation, the radiant gas cooler reduces exit raw gas temperature to 1,250 °F. The capacity of a single gasifier in the reference case is on the order of 2,200 tons/day coal.

Exiting the gasifier, raw fuel gas is scrubbed with water to remove particulates. Water is separated from the slag, and flows to the sour water stripper for treatment. Raw fuel gas mixes with steam for COS hydrolysis and two-stage water gas shift. Heat recovered from the high temperature shift reactor is recovered to generate high pressure steam. Heat recovered from the low temperature gas shift is suitable for generating intermediate pressure steam. The feed rate of shift steam is regulated in order to shift CO in the raw fuel gas sufficient to meet 90 % carbon removal overall.

Following the shift reaction, the gas is cooled again; first to 315 °F to recover useful heat for low pressure steam generation, next to 235 °F to recover useful heat for the steam cycle deaerator, then finally to 100 °F for NH3 removal. The cooling temperatures of 315 °F and 235 °F were selected based on reasonable temperature approaches to the steam cycle streams.

The fuel gas enters packed carbon bed absorbers to remove mercury, followed by a two-stage Selexol process that absorbs both CO_2 and H_2S from the fuel gas. H_2S is stripped from the solvent in the solvent regenerator and sent to the Claus plant. The CO_2 is compressed to 2,200 psig for transport to sequestration.

The Claus plant converts H_2S to elemental sulfur through a series of reactions. Sulfur is condensed, and tail gas is hydrogenated to convert residual SO_2 back into H_2S, which can be captured when the tail gas is recycled to the Selexol absorber. A small slipstream of clean fuel gas is used for reactant.

Clean fuel gas exits the Selexol absorber at nearly 700 psia, and is delivered to the topping combustor at 464.7 psia. Therefore, it can be expanded to recover excess pressure prior to entering the topping combustor; this expansion results in about 6 MWe of power generation.

Fuel gas is diluted with N_2 from the ASU; the hydrogen-rich mixture is burned in the topping combustor. Because of the high H_2 content, the fuel flowrate is regulated to maintain a turbine exit temperature of 1,050 °F. The net turbine power output is 192 MWe per unit [3].

All available process heat is collected for steam generation in the bottoming cycle. Superheated steam is expanded through three turbines, with reheat after the high pressure turbine. The steam cycle also provides heat to generate shift steam, acid gas removal (the Selexol solvent regenerator), the sour water stripper, and fuel gas reheating prior to the fuel gas expander.

2.2. Advanced Technology Assumptions

In the absence of demonstration data, process performance and costs for unproven futuristic technologies are difficult to estimate. Engineering judgment and information provided by technology developers are used, when necessary, to derive reasonable estimates. In addition, performance and cost results are provided to technology developers for reasonableness review and comment. While every attempt is made to calculate objective and reasonable performance and cost results, the bottom-line accuracy is limited by the uncertainty of design information.

At the time that the cases were configured, the limitations and key assumptions were as follows.

Technology Advancement	Performance Limitations and Assumptions	Cost Limitations and Assumptions
Advanced H_2 Turbines	Turbine parameters are highly proprietary to technology developers, and detailed turbine simulation modeling is outside the scope of this study. Hydrogen turbine parameters are devised to a configuration that meets the turbine program goals of 3- 5 percentage points above a 7FA turbine. Technology developers have performed system analyses using proprietary data and advanced modeling that predict their R&D efforts will exceed this goal. Performance uncertainty also exists due to limited commercial experience with hydrogen-fired turbines.	Turbine cost is scaled to the turbine power rating. There is no assumed premium for additional cost at elevated temperature or pressure. Increases in turbine power ratings result in plant-wide economies of scale resulting from increased net plant power production. For this reason, capital costs and COE are sensitive to the assumed turbine power rating increase and the scaling factors used on all plant equipment.
Coal Feed Pump	The coal feed pump is assumed to process as-received coal – without the need for coal drying. Demonstration to 1,000 psia pressure has been verified.	While there is considerable uncertainty regarding the cost of the coal feed pump, it is expected to be a relatively small capital cost which, when divided by the net plant power output to calculate on a $/kW basis, will have a minor impact on COE.
Warm Gas Cleanup	Extents of reaction and pressure drop through vessels are based on technology description by the developer. A demonstration scale unit has been running at the Eastman gasifier in Kingsport Tennessee but data from that demonstration has not yet been incorporated into this model. Reports on that demonstration plant indicate that the technology is performing well with very low exit concentrations of sulfur.	Technology developer's target costs are utilized in cost assessments. Installation costs, EPC costs and process and project contingencies are added as appropriate.
Hydrogen Membrane	The DOE/N ETL Hydrogen and Clean Fuels Program 2015 target flux and temperatures are used in simulating performance. Commercialization of high temperature hydrogen membranes must surmount challenges of (1) manufacturing membranes with consistent high flux properties and long lifetimes, and (2) fabrication of the membrane units themselves with gas inlet and outlet interconnects.	2015 target membrane costs from the DOE/N ETL Hydrogen and Clean Fuels Program are utilized in cost assessments. Installation costs, EPC costs, and process and project contingencies are added as appropriate.

(Continued)

Technology Advancement	Performance Limitations and Assumptions	Cost Limitations and Assumptions
Ion Transport Membrane	Technology developer's target operational parameters such as pressure and flux are utilized in process simulations. Very promising results have been obtained in the 5TPD oxygen demo unit that is operating at the Sparrows Point refinery in Maryland.	Technology developer's target costs are utilized in cost assessments. Installation costs, EPC costs and process and project contingencies are added as appropriate.
Reliability, Availability and Maintainability (RAM)	R&D in areas improving RAM may impact process performance; however, for this analysis, any changes in process efficiency are assumed to be negligible.	Improved RAM is modeled by increasing the capacity factor from 80% to 85% to 90%. This study does not specifically tie DOE-funded projects to capacity factor improvements. Capital costs associated with improved RAM are assumed to be negligible.
Solid Oxide Fuel Cell	The IGFC configuration includes the following: (1) an advanced pressurized SOFC meeting DOE/NETL Fuel Cell Program performance targets; (2) a conceptual catalytic gasifier that provides high methane content syngas, and (3) a pressurized oxycombustor that burns the hot spent anode fuel gas from the SOFC. Heat generated in the SOFC can be partially dissipated by internally reforming methane in the syngas. The catalytic gasifier is conceptual and is assumed to produce 17 mole % CH_4 by the potassium catalyzed methanation reaction. This is exothermic and helps to drive the endothermic gasification reaction. Great Point Energy is developing a catalytic gasifier that is based on the original Exxon process whereby the methanation reaction can provide enough heat for gasification so that oxygen is not required.	The fuel cell system total plant costs are assumed to be $700/kW (gross power from the fuel cell). Stack replacement frequency and cost are based on DOE/NETL Fuel Cell Program targets. The catalytic gasification costs are assumed to be based on the same reference costs as the non-catalytic gasification systems and scaled on coal throughput. Catalyst recovery costs are included.

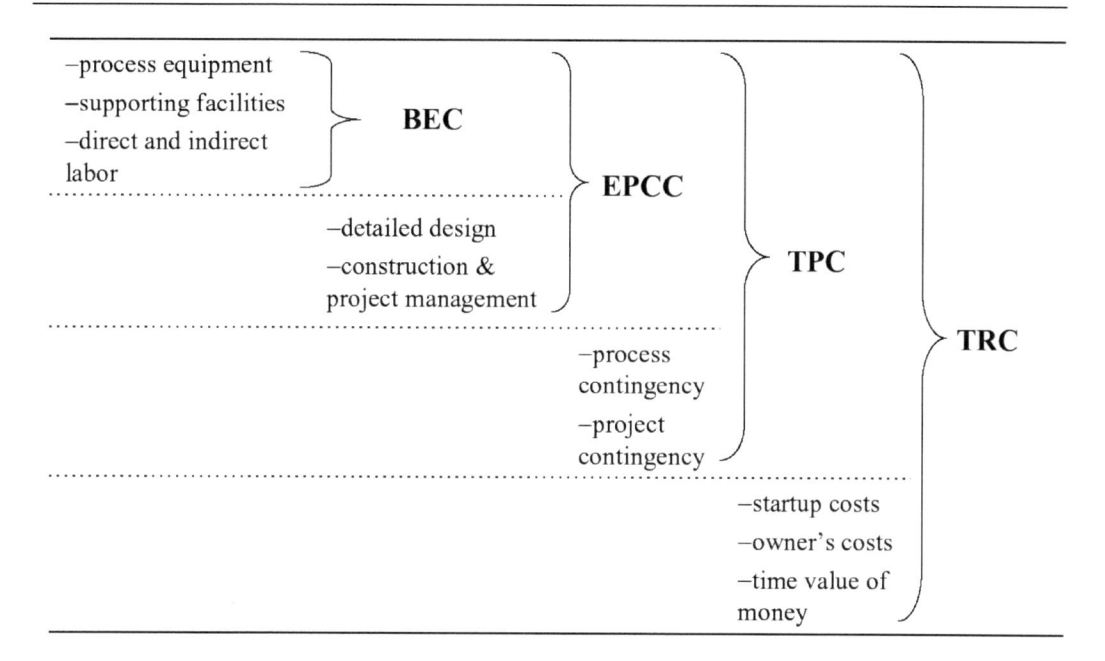

Figure 2-2. Elements of Capital Cost.

2.3. Economic Analysis

See Appendix A for NETL's update to capital costs and COE. [3]

Plant capital cost is estimated using cost algorithms based on literature and vendor supplied costs and capacities consistent with this level of conceptual scope definition and taking into consideration plant size, number of process trains, sparing philosophy, and as much equipment- specific design information as possible.

Operating and maintenance (O&M) costs include fixed labor costs as well as variable costs (that depend on capacity factor) including maintenance materials, water, chemicals, and waste disposal. Fuel cost is calculated separately from O&M based on coal feed rate and coal cost.

The cost of electricity calculation (described below) can be based directly on the capital charge factor. This study assumes a prescribed capital charge factor (17.5 %) typical of a higher-risk project undertaken by an investor-owned utility.

2.3.1. Capital Cost

The following Figure 2-2 illustrates the relationships between various elements of capital cost. Noblis correlations are used to estimate Bare Erected Cost (BEC) for each major section of the process plant. The BEC is estimated (in January 2007 dollars) using mass and energy balance information from Aspen Plus simulations of each case. For ease in comparing results, the organization of plant sections is consistent with the presentation used in NETL's Baseline Study. Each section's BEC represents the sum of major plant equipment within the section (including initial chemical and catalyst loadings), as well as materials and labor. Appropriate for a scoping study, BEC's are based on scaled estimates using best-available information collected from multiple sources for the cost correlations.

The BEC is used as the basis for calculating detailed engineering and construction and project management fees. A 9 % charge is applied which, when added to the BEC, becomes the Engineering, Procurement, and Construction Cost (EPCC). The cost analyses in Chapter 3 of this report present the Total Plant Cost (TPC) at the process section level; however the capital cost contains additional process section detail for BEC, EPCC, and process and project contingencies.

For consistency, process and project contingencies used in NETL's Baseline Study form the basis for all major equipment in each plant section. Advanced technologies are assumed to embed cost uncertainty in the BEC; in this manner they retain the same level of contingency as conventional technologies in order not to put the advanced technologies at a disadvantage due to contingency. Contingency estimates are added to the EPCC to calculate the TPC.

Startup costs (assumed to be 2 % of EPCC), owner's costs (which might typically include a Technology Fee or licensing fee), and the time value of money are normally added to the TPC in order to obtain the Total Required Capital (TRC). For consistency with NETL's Baseline Study, owner's costs are omitted in this economic analysis because they are project-specific. Therefore, the reader should bear in mind that the financial results of this analysis (levelized cost of electricity and capital charge factor) do not include owner's costs.

Table 2-2. Elements of Variable Operating Cost

Maintenance Materials
Water
Chemicals
Carbon (Hg removal)
COS Catalyst
Shift Catalyst
Claus Catalyst
Selexol Solvent
ZnO Sorbent
Membrane Replacement
Fuel Cell Stack Replacement
Spent Catalyst Waste Disposal
Ash Disposal

2.3.2. O&M Cost

Labor represents a fixed operating cost, and is based on the number of operating laborers in the plant. The Baseline Study estimate for number of laborers, labor rates, burden, and administrative overhead is used as a basis. Administrative labor is estimated as an overhead rate (25 %) to the sum of operating and maintenance labor. An average labor rate of $33/hr is assumed – again consistent with that used in NETL's Baseline Study.

Table 2-2 identifies elements of variable operating cost that are included in the analysis. Consistent with the Baseline Study, no credit is taken for by-products from any process.

Fuel cost is calculated based on net power generation, heat rate, and fuel heating value. A coal cost of $42.11/ton ($1.80/MMBtu) is assumed, with an as-received heating value of 11,666 Btu/lb. For warm gas cleanup, costs of $14,000/ton for ZnO sorbent and $100/ton for trona are assumed[4]. The sorbent attrition rate is assumed to be 10-20 lb. per million lb. circulating sorbent.

2.3.3 Cost of Electricity

The current-dollar levelized cost of electricity can be calculated using the formula:

$$COE_P = ((CCF_P * TPC) + LF_{FP} * FYC_F + CF * (LF_{1P} * FYC_1 + LF_{2P} * FYC_2 + \ldots)) / (CF * MWh) + TSM$$

Where:

COE_P = levelized cost of electricity over P years
CCF_P = capital charge factor levelized over P years
TPC = total plant cost
LF_{FP} = levelization factor over P years for fixed operating costs
FYC_F = first year fixed operating costs
CF = capacity factor
LF_{nP} = levelization factor over P years for category n variable operating cost element
FYC_n = first year variable operating costs for category n cost element
MW_h = net annual power generation at 100% capacity factor
TSM = charge for CO_2 transportation, storage, and monitoring

The capital charge factor can be considered to be the rate at which capital costs are recovered during the lifetime of the project. It is a function of cost of capital and level of technology risk; as these factors increase, the capital charge factor also increases. For the purposes of this study, the investment scenario is considered to be an investor-owned utility (IOU) involved in higher- risk technology. Consistent with NETL's Baseline Study, the capital charge factor in this scenario is 17.5 %. Additional assumed financial parameters are itemized in Table 2-3.

Individual levelization factors for the COE equation above can be calculated by:

$$LF_{nP} = k * (1 - k^P) / (a_P * (1 - k))$$

Where

k = $(1 + e) / (1 + i)$
a_P = $(((1 + i)^P - 1) / (i * (1 + i)^P)$
e = annual escalation rate
i = annual discount rate

Consistent with NETL's Baseline Study, the 20-year O&M levelization factors for both fixed and variable costs are 1.1568 (presumes an escalation rate of 1.87 %). For coal, the 20-year levelization factor is 1.2022 (presumes an escalation rate of 2.35 %). Once again, all costs in this analysis are based on January 2007 dollars.

Table 2-3. Discounted Cash Flow Analysis Parameters

Parameter	Value
Percentage Debt	45 %
Interest Rate	11.55 %
Repayment Term of Debt	15 years
Grace Period on Debt Repayment	0 years
Debt Reserve Fund	None
Depreciation	20 years; 150 % DB
Working Capital	Zero
Plant Economic Life	30 years
Coal Escalation Factor	2.35 %
O&M Escalation Factors	1.87 %
EPC Escalation	0 %
Tax Holiday	0 years
Income Tax Rate	38 %
Investment Tax Credit	0 %
Duration of Construction	36 months

Table 3-1. Carbon Capture Power System Technology Development

Case Title	Gas Turbine	Coal Feed System / Gasifier	Capacity Factor	Gas Clean Up	CO_2 Separation	Oxygen Production
Reference IGCC	7FA	Slurry Feed	80% CF	2-Stage Selexol		Cryogenic
Adv "F" Turbine	Adv "F"					Air
Coal Feed Pump		Coal				Separation
85% CF		Feed	85% CF			Unit
WGCU/Selexol		Pump		WGCU	Selexol	(ASU)
WGCU/ H_2Membrane					High	
AHT-1 Turbine	AHT-1				Temp	
ITM					Hydrogen	ITM
AHT-2 Turbine	AHT-2				Membrane	
90% CF			90% CF			
Advanced IGFC	Pressurized SOFC	Catalytic Gasifier	90% CF	WGCU	SOFC + Oxycom-bustion	Cryogenic ASU

Finally, a CO_2 transmission, storage, and monitoring (TS&M) charge of 3.9 mills/kW-hr is applied to the COE to account for CO_2 sequestration.

3. ANALYSIS OF ADVANCED POWER PROCESS CONFIGURATIONS WITH CARBON CAPTURE

A series of process configurations with carbon capture that produce electric power from bituminous coal is analyzed to determine the potential performance improvements and cost reductions resulting from advanced technology. Starting with the reference IGCC plant with carbon capture, process modifications are simulated to represent commercialization of advanced technologies. These process configurations are listed in Table 3-1. The white blocks represent existing, commercially available technologies while the colored blocks represent advanced emerging technologies. Each advanced technology is implemented and evaluated in a composite process and in the order in which demonstration-readiness is anticipated. This allows assessment of the cumulative improvements in process performance and cost over time. The majority of the technologies are evaluated in the context of an IGCC plant. The pressurized SOFC case represents an advanced process configuration later in the demonstration timeline, incorporating some technologies that are of specific value to an integrated gasification fuel cell (IGFC) plant.

3.1. Carbon Capture Reference Plant

The process configurations used for both the capture and non-capture reference plants are based on state-of-the-art technology available in 2003 – the basis DOE used to establish its R&D program goals. The carbon capture reference plant is the same IGCC process as the non-capture reference plant from Volume 1 of this pathway study, except that the gas cleanup section has a sour shift to produce H_2-rich fuel and CO_2. The CO_2 is separated and compressed for pipeline transport to long-term storage; the H_2-rich fuel powers the hydrogen turbine. All IGCC carbon capture technologies in this study are based on 90 % capture of the carbon derived from coal.

Case Configuration: Slurry Feed Gasifier, Cryogenic ASU, Cold Gas Cleanup, 7FA Hydrogen Turbine, 80 % Capacity Factor

The carbon capture reference plant includes slurry feed gasifier, cryogenic air separation, cold gas cleanup, 7FA-based hydrogen turbine, CO_2 compression, and 80 % capacity factor. Water gas shift and CO_2 separation (achieved using 2-stage Selexol) are included as part of the gas cleanup section.

Figure 3-1 presents a block flow diagram of the process. Colored boxes in the illustration indicate process sections that are different from the non-capture reference process. The plant is configured with the following:

- Two trains of single-stage slurry feed gasifiers with radiant-only syngas coolers
- Two cryogenic air separation units
- Two trains of water quench and sour water gas shift/carbonyl sulfide (COS) hydrolysis
- Two trains of 2-stage Selexol acid gas removal
- Four trains of CO_2 compressors

- One train of sulfur recovery using conventional Claus technology
- Two trains of 7FA hydrogen turbines
- One HRSG
- One steam turbine bottoming cycle with high, intermediate, and low pressure (condensing) turbine sections; steam conditions are 1,800 psi and 1,000 °F for the high pressure turbine and 405 psi and 1,000 °F for the intermediate pressure turbine.

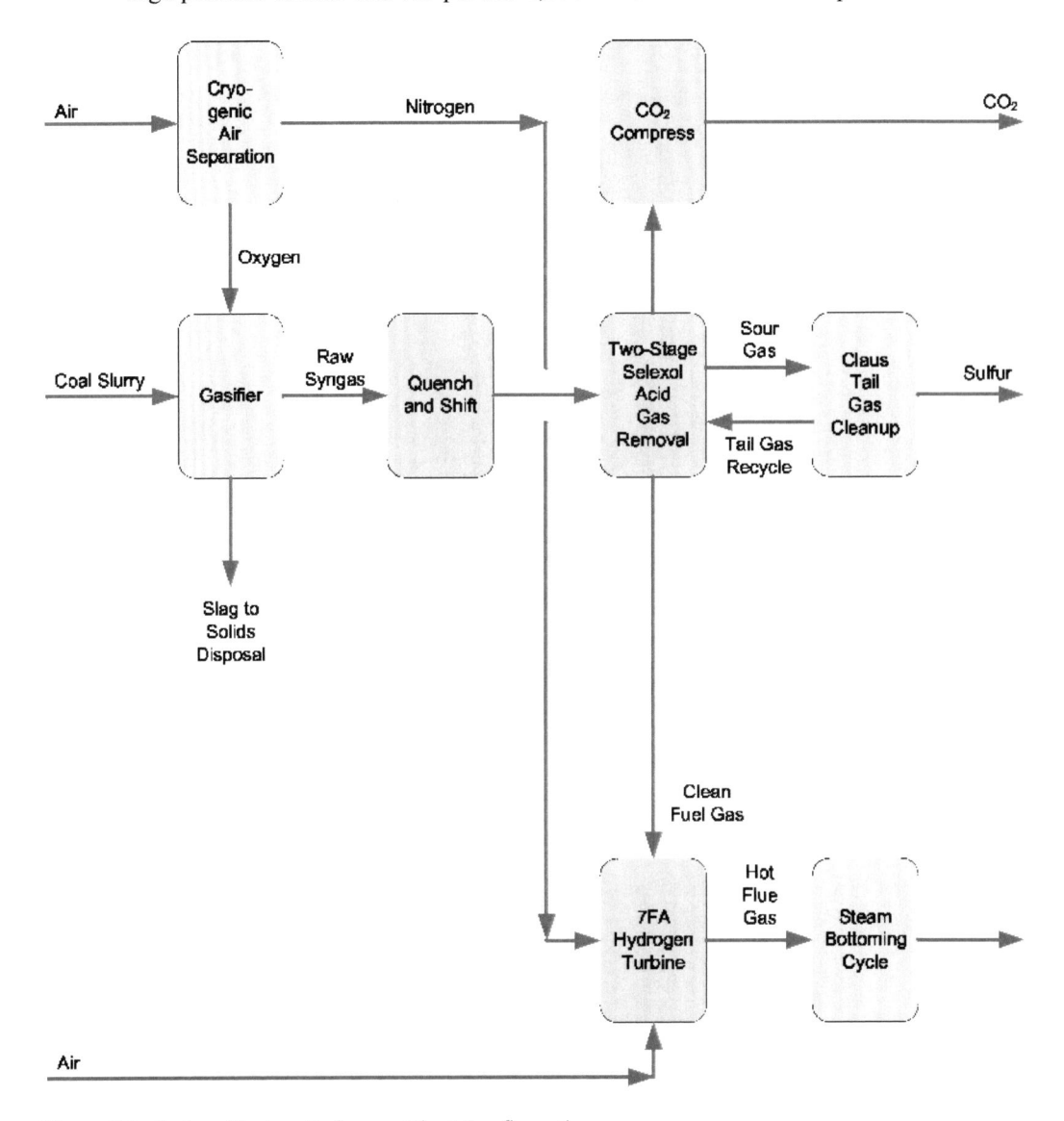

Figure 3-1. Carbon Capture Reference Plant Configuration.

This IGCC plant produces a net 444 MW of power. Carbon utilization is 98 %, and overall efficiency is 30.4 % (HHV basis). Comparison with the non-capture reference plant in Table 3-2 illustrates the differences in process performance resulting from carbon capture. The same turbine size and power rating are assumed for syngas and hydrogen fuel.[5] The smaller heating value per mole of H_2 in hydrogen fuel compared to CO in syngas fuel results in a greater coal requirement for the carbon capture case; the additional heat recovery available due to this increased coal feed rate more than counters the shift steam requirement associated with the capture configuration, resulting in an increase in steam turbine power generation of 14 MW.

Table 3-2. Performance Impact of Carbon Capture in the Reference Plant

	Non-Capture Reference Plant	Carbon Capture Reference Plant
Gas Turbine Power (MWe)	384	384
Fuel Gas Expander (MWe)	6	6
Steam Turbine Power (MWe)	223	237
Total Power Produced (MWe)	614	627
Auxiliary Power Use (MWe)	-127	-183
Net Power (MWe)	487	444
As-Received Coal Feed (lb/hr)	402,581	426,544
Net Heat Rate (Btu/kW-hr)	9,649	11,214
Net Plant Efficiency (HHV)	35.4 %	30.4 %

Auxiliary power use increases by 56 MW in the carbon capture case due to (1) increased plant size in general because of increased coal feed rate, (2) addition of CO_2 compressors, and (3) increased Selexol auxiliary power as the result of separating both H_2S and CO_2. In the reference plant, therefore, CO_2 capture imposes a 5.0 percentage point decrease in process efficiency from the non-capture case.

Cost Analysis

See Appendix A for NETL 's update to capital cost and COE.

Table 3-3 below compares the Total Plant Cost (TPC) for major sections of each process plant.

TPC increases by roughly between 2 to 5 % for most plant sections due to the increase in coal feed rate and therefore generally larger plant size in the carbon capture case. TPC on a $/kW basis, however, increases by a higher percentage (typically between 12 to 14 %) as the result of 43 MW less net power generation from the carbon capture case.

Gas cleanup section cost increases by a factor of about 2 due to (1) additional cost of water gas shift reactors (not used in the non-capture process), and (2) cost of the additional Selexol stage for CO_2 separation in the carbon capture case. The CO_2 compression section is an additional $94/kW cost to the carbon capture case that is not present in the non-capture plant. The cost of the hydrogen turbine is assumed to increase slightly in the carbon capture cases as the result of modifications required for H_2-rich fuel as opposed to syngas fuel.

Labor cost increases in the carbon capture case due to (1) slightly greater plant size resulting from increased coal feed rate, and (2) increased plant complexity from additional water gas shift, two-stage Selexol, and CO_2 compression sections.

Variable operating costs are calculated based on 80 % capacity factor. Results from the cost analysis indicate a TPC of $2,718/kW and a 20-year levelized COE of $0.1148/kW-hr based on January 2007 dollars. Compared to the non-capture plant, these represent a 29 % increase in both TPC ($/kW basis) and in COE due to CO_2 capture and storage.

Table 3-3. Reference Plant Capital and O&M Cost Comparison

	Non-Capture Reference Plant		Carbon Capture Reference Plant		Δ	
Capital Cost ($1,000)						
Plant Sections	TPC	TPC $/kW	TPC	TPC $/kW	Δ TPC $/kW	% Δ
1 Coal and Sorbent Handling	30,821	63	31,944	72	9	14
2 Coal and Sorbent Prep & Feed	48,980	101	50,928	115	14	14
3 Feedwater & Balance of Plant	35,077	72	36,260	82	10	14
4a Gasifier	236,212	485	241,531	544	59	12
4b Air Separation Unit	168,950	347	175,776	396	49	14
5a Gas Cleanup	112,389	231	206,045	464	233	101
5b CO_2 Removal & Compression	0	0	41,703	94	94	∞
6 Gas Turbine	105,058	215	116,181	262	47	22
7 HRSG	49,511	102	48,250	109	7	7
8 Steam Cycle and Turbines	54,310	112	56,734	128	16	14
9 Cooling Water System	24.233	50	25,010	56	6	12
10 Waste Solids Handling System	38,752	80	40,159	91	11	14
11 Accessory Electric Plant	66,529	137	73,922	167	30	22
12 Instrumentation & Control	23,178	48	25,730	58	10	21
13 Site Preparation	18,143	37	18,780	42	5	14
14 Buildings and Structures	16,314	34	16,931	38	4	12
Total	1,028,457	2,113	1,205,882	2,718	605	29
O&M Cost ($1,000/yr)						
Fixed Costs	Total		Total		Δ	% Δ
Labor	19,542		22,548		3,006	15
Variable Operating Costs*	Total		Total			
Maintenance Materials	19,593		21,569		1,976	10
Water	1,548		1,732		184	12
Chemicals	1,089		1,838		749	69
Waste Disposal	2,413		2,560		147	6
Total Variable Costs	24,642		27,698		3,056	12
Total O&M Cost	44,184		50,247		6,063	14
Fuel Cost*	59,402		62,938		3,536	6
Discounted Cash Flow Results, levelized						
Capital Cost ($/kW-hr)	0.0528		0.0679			29
Fixed O&M Cost ($/kW-hr)	0.0066		0.0084			27
Variable O&M Cost ($/kW-hr)	0.0084		0.01 03			23
Fuel Cost ($/kW-hr)	0.0209		0.0243			16
TS&M Cost ($/kW-hr)	0		0.0039			∞
Levelized COE ($/kW-hr)	0.0887		0.1148			29

*Includes 80 % Capacity Factor

3.2. Advanced "F" Frame Hydrogen Turbine

The advanced "F" hydrogen turbine produces more power, has a higher pressure ratio, and a higher firing temperature than the 7FA-based hydrogen turbine. Turbine performance is based on the carbon capture IGCC case in NETL's Baseline Study.

In non-capture cases, three benefits associated with the advanced "F" syngas turbine are (1) integration with the ASU reduces the auxiliary power load of the ASU (a portion of the air supply to the ASU is provided by the gas turbine), (2) the higher turbine firing temperature results in improved turbine performance, and (3) subsequently higher turbine exhaust temperature allows an increase in the steam superheat temperature from 1,000 °F to 1,050 °F.

In the carbon capture cases, these benefits are significantly diminished because (1) no air is extracted from the hydrogen turbine because there would not be sufficient flow through the turbine to meet both its power rating and operating temperature specifications, (2) turbine firing temperature is limited (due to the high moisture content in the turbine exhaust) by materials limitations, and (3) limited exhaust temperature of 1,050 °F provides a temperature differential for steam superheat temperature no higher than 1,000 °F.

Case Configuration: Slurry Feed Gasifier, Cryogenic ASU, Cold Gas Cleanup, Advanced "F" Frame Hydrogen Turbine, 80 % Capacity Factor

A block flow diagram of this case is presented in Figure 3-2. This two-train IGCC plant produces a net 539 MW of power. Overall efficiency is 31.7 % (HHV basis). Carbon utilization is 98 % and the capacity factor is 80 %. Performance resulting from the advanced "F" hydrogen turbine is compared against the 7FA turbine case in the following Table 3-4.

Table 3-4. Incremental Performance Improvement from Advanced "F" Hydrogen Turbine

	Carbon Capture Reference Plant	Advanced "F" Turbine
Gas Turbine Power (MWe)	384	464
Fuel Gas Expander (MWe)	6	7
Steam Turbine Power (MWe)	237	274
Total Power Produced (MWe)	627	745
Auxiliary Power Use (MWe)	-183	-206
Net Power (MWe)	444	539
As-Received Coal Feed (lb/hr)	426,544	496,865
Net Heat Rate (Btu/kW-hr)	11,214	10,755
Net Plant Efficiency (HHV)	30.4 %	31.7 %

The 7FA-based hydrogen turbine in the reference case is rated at 192 MW, while the advanced "F" turbine is rated at 232 MW. Because of the increased coal feed rate needed to power the higher-rated turbine, steam turbine power generation and auxiliary power use increase.

The increased power rating and pressure ratio of the advanced "F" hydrogen turbine result in a 1.3 percentage point efficiency improvement in the carbon capture cases. In the corresponding non-capture assessment, process efficiency increases by 2.5 percentage points.

As discussed above, factors that limit performance efficiency improvement in this carbon capture case are: (1) the absence of air integration results in relatively greater ASU auxiliary load relative to coal feed rate, (2) steam turbine power increases by only 40 MW (as opposed to a 70 MW increase in the non-capture analysis) because of the limited turbine firing temperature that results in less sensible heat carried through the HRSG by the flue gas, and (3) lower steam superheat temperature that reduces the Carnot efficiency of the steam cycle below that achieved in the non- capture cases.

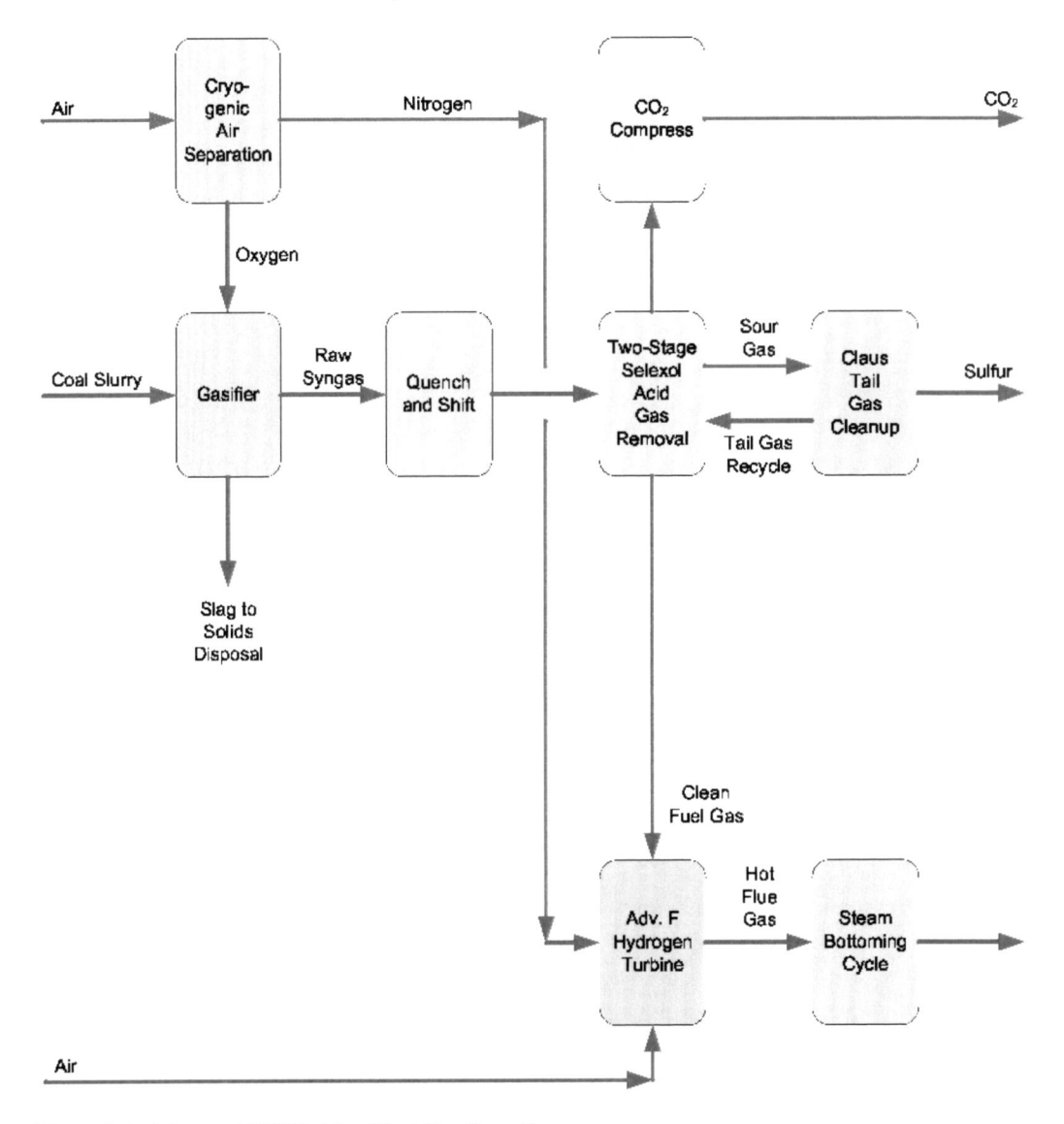

Figure 3-2. Advanced "F" Turbine Plant Configuration.

Table 3-5. Advanced "F" Turbine: Capital and O&M Cost Comparison

	Carbon Capture Reference Plant		Advanced "F" Turbine		Δ	
Capital Cost ($1,000)						
Plant Sections	TPC	TPC $/kW	TPC	TPC $/kW	Δ TPC $/kW	% Δ
1 Coal and Sorbent Handling	31,944	72	35,118	65	-7	-10
2 Coal and Sorbent Prep & Feed	50,928	115	56,449	105	-10	-9
3 Feedwater & Balance of Plant	36,260	82	38,079	71	-11	-13
4a Gasifier	241,531	544	266,942	495	-49	-9
4b Air Separation Unit	175,776	396	194,517	361	-35	-9
5a Gas Cleanup	206,045	464	230,927	428	-36	-8
5b CO_2 Removal & Compression	41,703	94	48,578	90	-4	-4
6 Gas Turbine	116,181	262	131,969	245	-17	-6
7 HRSG	48,250	109	53,454	99	-10	-9
8 Steam Cycle and Turbines	56,734	128	62,886	117	-11	-9
9 Cooling Water System	25,010	56	26,771	50	-6	-11
10 Waste Solids Handling System	40,159	91	44,115	82	-9	-10
11 Accessory Electric Plant	73,922	167	78,735	146	-21	-13
12 Instrumentation & Control	25,730	58	26,588	49	-9	-16
13 Site Preparation	18,780	42	19,241	36	-6	-14
14 Buildings and Structures	16,931	38	17,615	33	-5	-13
Total	1,205,882	2,718	1,331,986	2,472	-246	-9
O&M Cost ($1,000/yr)						
Fixed Costs	**Total**		**Total**		Δ	% Δ
Labor	22,548		25,555		7	0
Variable Operating Costs*	**Total**		**Total**			
Maintenance Materials	21,569		24,357		2,788	13
Water	1,732		1,885		153	9
Chemicals	1,838		2,115		277	15
Waste Disposal	2,560		2,965		405	16
Total Variable Costs	27,698		31,322		3,624	13
Total O&M Cost	50,247		56,877		6,630	13
Fuel Cost*	62,938		73,314		10,376	16
Discounted Cash Flow Results, levelized						
Capital Cost ($/kW-hr)	0.0679		0.0617			-9
Fixed O&M Cost ($/kW-hr)	0.0084		0.0078			-7
Variable O&M Cost ($/kW-hr)	0.01 03		0.0096			-7
Fuel Cost ($/kW-hr)	0.0243		0.0233			-4
TS&M Cost ($/kW-hr)	0.0039		0.0039			0
Levelized COE ($/kW-hr)	0.1148		0.1064			-7

*Includes 80 % Capacity Factor

Cost Analysis

See Appendix A for NETL 's update to capital cost and COE.

Table 3-5 below compares capital and O&M costs with the carbon capture reference plant. The change in gas turbine drives the differences in capital costs between the reference plant and the case with advanced "F" hydrogen turbine. The advanced "F" turbine has a higher power rating, which increases coal flowrate to the process, and therefore larger equipment sizes throughout the plant; this is reflected in the higher TPC costs in the advanced "F" case. On a $/kW basis, however, the TPC of the advanced "F" turbine plant decreases by about 9 % because of increased net power output.

When the advanced "F" turbine is incorporated into the non-capture cases, TPC decreases by about 17 % (on a $/kW basis); the relative reduction in TPC is somewhat less for the carbon capture cases (9 %). Three primary reasons for this are (1) the cost of the main air compressor increases (rather than decreases) because there is no air integration in the advanced "F" turbine carbon capture case, (2) the bottom-line TPC is greater for the capture cases (because of greater coal throughput than the non-capture cases and also the additional cost for shift, two-stage Selexol, and CO_2 removal and compression) so the percentage decrease in TPC is more difficult to attain, and (3) the incremental net power generated in the carbon capture cases (95 MW) is less than the non-capture cases (150 MW) which results in less of a decrease in TPC on a $/kW basis. The advanced "F" hydrogen turbine in the carbon capture cases results in a smaller percentage decrease in TPC on a $/kW basis than in the non-capture cases.

Corresponding with the 9 % decrease in TPC (on a $/kW basis) from the carbon capture reference plant to the advanced "F" turbine plant, the COE decreases by about 7 % – from $0.1148/kW-hr to $0.1064/kW-hr. That result is based on 80 % capacity factor. The decrease in COE between the carbon capture cases is less than in the non-capture cases for all the same reasons as the TPC.

3.3. Coal Feed Pump

The coal feed pump replaces the slurry feed system, delivering as-received coal to the gasifier which eliminates the energy required to evaporate slurry water in the gasifier thereby increasing cold gas efficiency of the gasifier.

Case Configuration: Coal Feed Pump, Cryogenic ASU, Cold Gas Cleanup, Advanced "F" Hydrogen Turbine, 80 % Capacity Factor

This process configuration, shown in Figure 3-3, is identical to that in Figure 3-2 except that as- received coal is delivered to the gasifier rather than coal slurry. Dry feed has the advantage of less energy consumed in the gasifier to evaporate water from the slurry, resulting in a greater portion of the coal feed converted to CO (rather than CO_2) in the raw syngas.

The raw syngas composition in this case has much less water than the previous case because of the dry feed. Due to the higher cold gas efficiency of the gasifier, less coal is needed in this case, so the molar flowrate of raw syngas is also less. The concentration of CO is much greater due to not having to oxidize as much carbon in the gasifier in order to evaporate slurry water. The absence of moisture from slurry water in the coal feed pump case also means that relatively more shift steam must be added.

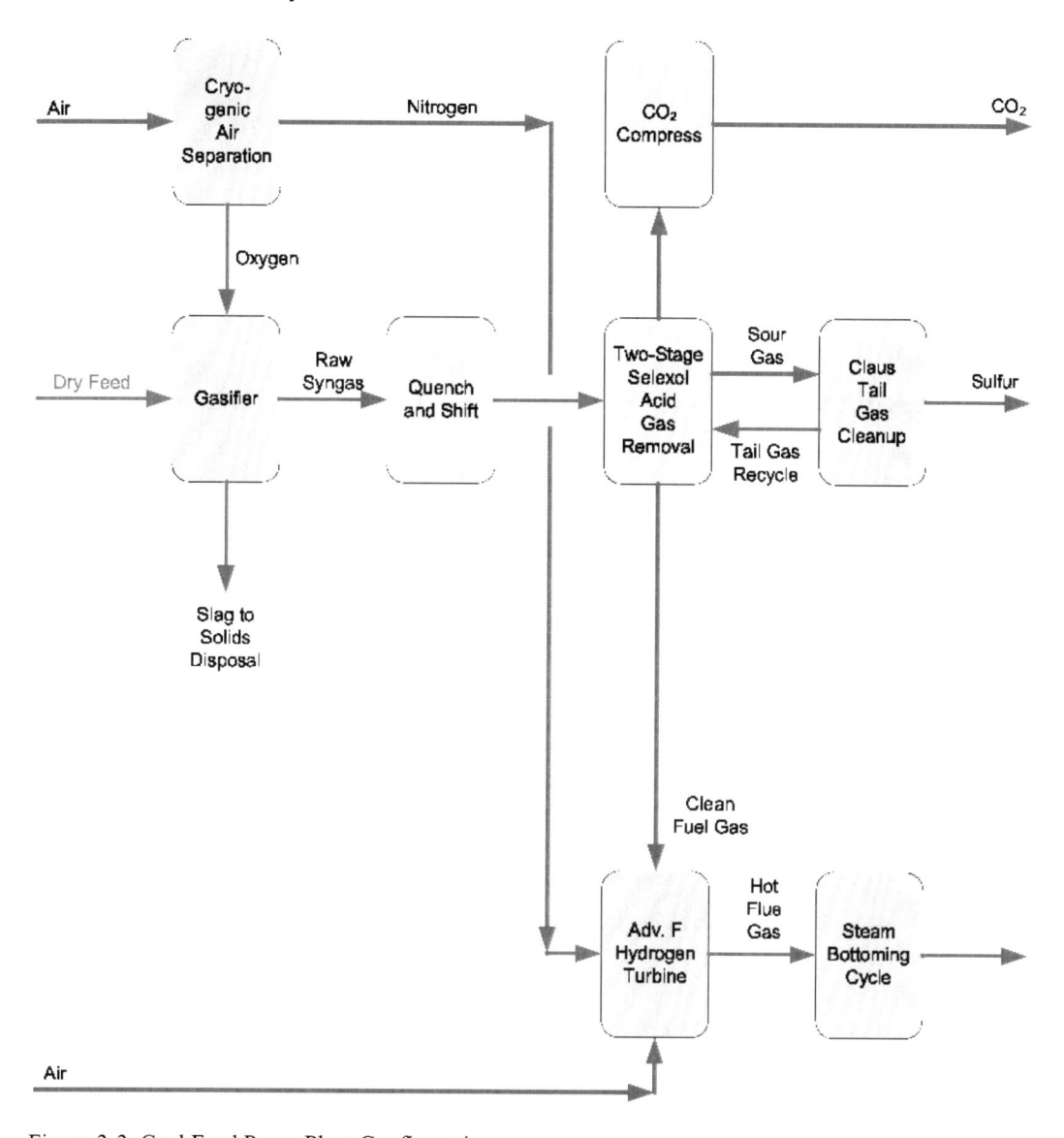

Figure 3-3. Coal Feed Pump Plant Configuration.

Table 3-6 illustrates the primary differences in process performance resulting from slurry feed versus dry feed gasifier operation. Total power production is 45 MW less in the coal feed pump case because of less power recovered by the steam cycle – due primarily to (1) less heat recovered by the gasifier radiant cooler and syngas cooling section as the result of decreased coal throughput and less molar flow because there is less water in the syngas, and (2) additional shift steam generation due to the lack of water in the coal feed. Auxiliary power consumption relative to the coal feed rate is essentially constant; the reduction in ASU parasitic load correlates to the drop in coal feed rate. Overall, the net power generated in the coal feed pump case is 29 MW less than the slurry feed case, but the coal feed rate required to achieve the 232 MWe gas turbine rating is also significantly lower – resulting in an improved net plant efficiency from 31.7 % to 32.5 %.

In the non-capture cases, the coal feed pump improves process efficiency by 2.1 percentage points. Compared to the 0.8 percentage point efficiency improvement for the carbon capture cases, the coal feed pump represents less of an improvement to the carbon capture cases because of the increase in shift steam that must be generated in the absence of slurry water.

Table 3-6. Incremental Performance Improvement from the Coal Feed Pump

	Advanced "F" Turbine	Coal Feed Pump
Gas Turbine Power (MWe)	464	464
Fuel Gas Expander (MWe)	7	7
Steam Turbine Power (MWe)	274	228
Total Power Produced (MWe)	744	699
Auxiliary Power Use (MWe)	-206	-189
Net Power (MWe)	539	510
As-Received Coal Feed (lb/hr)	496,865	459,257
Net Heat Rate (Btu/kW-hr)	10,755	10,497
Net Plant Efficiency (HHV)	31.7 %	32.5 %
Gasifier Cold Gas Efficiency	76.1 %	81.9 %

Cost Analysis

See Appendix A for NETL 's update to capital cost and COE.

Capital and O&M costs are compared with the slurry feed case results in Table 3-7. Total plant cost generally decreases in the coal feed pump case due to less coal feed rate, and therefore smaller equipment sizes. The cost per kilowatt remains about the same in most cost accounts, however, because of decreased power production.

The gas turbine and H RSG absolute costs do not change between cases because these remain the same size due to the fixed power output of the advanced "F" turbine; however, the costs on a $/kW basis increase for these plant sections in the coal feed pump case due to the decreased net power output.

The $74 MM reduction in TPC from the slurry feed case to the dry feed case is almost the same as the $80 MM reduction in the non-capture cases. However, decreased power production in the carbon capture cases results in only a $7/kW reduction in TPC compared to the $60/kW reduction in the non-capture cases. The capital cost advantage of the coal feed pump is not as great in the carbon capture scenario as it is in the non-capture scenario.

The slight change in TPC for the carbon capture coal feed pump case translates to a slight reduction in COE from $0.1064/kW-hr to $0.1054/kW-hr – a 1.0 % decrease in COE.

Table 3-7. Coal Feed Pump: Capital and O&M Cost Comparison

	Advanced "F" Turbine		Coal Feed Pump		Δ	
Capital Cost ($1,000)						
Plant Sections	**TPC**	**TPC $/kW**	**TPC**	**TPC $/kW**	**Δ TPC $/kW**	**% Δ**
1 Coal and Sorbent Handling	35,118	65	33,445	66	1	2
2 Coal and Sorbent Prep & Feed	56,449	105	55,442	109	4	4
3 Feedwater & Balance of Plant	38,079	71	34,231	67	-4	-6
4a Gasifier	266,942	495	247,284	485	-10	-2
4b Air Separation Unit	194,517	361	173,695	340	-21	-6
5a Gas Cleanup	230,927	428	226,119	443	15	4
5b CO_2 Removal & Compression	48,578	90	45,607	89	-1	-1
6 Gas Turbine	131,969	245	132,079	259	14	6
7 HRSG	53,454	99	53,439	105	6	6
8 Steam Cycle and Turbines	62,886	117	55,118	108	-9	-8
9 Cooling Water System	26,771	50	24,402	48	-2	-4
10 Waste Solids Handling System	44,115	82	39,732	78	-4	-5
11 Accessory Electric Plant	78,735	146	75,981	149	3	2
12 Instrumentation & Control	26,588	49	25,937	51	2	4
13 Site Preparation	19,241	36	18,958	37	1	3
14 Buildings and Structures	17,615	33	16,627	33	0	0
Total	1,331,986	2,472	1,258,097	2,465	-7	0
O&M Cost ($1,000/yr)						
Fixed Costs	**Total**		**Total**		**Δ**	**% Δ**
Labor	25,555		24,051		-1,504	-6
Variable Operating Costs*	Total		Total			
Maintenance Materials	24,357		23,273		-1,084	-4
Water	1,885		1,434		-451	-24
Chemicals	2,115		1,969		-146	-7
Waste Disposal	2,965		2,502		-463	-16
Total Variable Costs	31,322		29,179		-2,143	-7
Total O&M Cost	56,877		53,230		-3,647	-6
Fuel Cost*	73,314		67,765		-5,549	-8
Discounted Cash Flow Results, levelized						
Capital Cost ($/kW-hr)	0.0617		0.0616			0
Fixed O&M Cost ($/kW-hr)	0.0078		0.0078			0
Variable O&M Cost ($/kW-hr)	0.0096		0.0094			-2
Fuel Cost ($/kW-hr)	0.0233		0.0228			-2
TS&M Cost ($/kW-hr)	0.0039		0.0039			0
Levelized COE ($/kW-hr)	0.1064		0.1054			-1

*Includes 80 % Capacity Factor

Table 3-8. 85 % Capacity Factor: Capital and O&M Cost Comparison

	Coal Feed Pump		85% Capacity Factor		Δ	
Capital Cost ($1,000)						
Plant Sections	**TPC**	**TPC $/kW**	**TPC**	**TPC $/kW**	**Δ TPC $/kW**	**% Δ**
1 Coal and Sorbent Handling	33,445	66	33,445	66	0	0
2 Coal and Sorbent Prep & Feed	55,442	109	55,442	109	0	0
3 Feedwater & Balance of Plant	34,231	67	34,231	67	0	0
4a Gasifier	247,284	485	247,284	485	0	0
4b Air Separation Unit	173,695	340	173,695	340	0	0
5a Gas Cleanup	226,119	443	226,119	443	0	0
5b CO_2 Removal & Compression	45,607	89	45,607	89	0	0
6 Gas Turbine	132,079	259	132,079	259	0	0
7 HRSG	53,439	105	53,439	105	0	0
8 Steam Cycle and Turbines	55,118	108	55,118	108	0	0
9 Cooling Water System	24,402	48	24,402	48	0	0
10 Waste Solids Handling System	39,732	78	39,732	78	0	0
11 Accessory Electric Plant	75,981	149	75,981	149	0	0
12 Instrumentation & Control	25,937	51	25,937	51	0	0
13 Site Preparation	18,958	37	18,958	37	0	0
14 Buildings and Structures	16,627	33	16,627	33	0	0
Total	1,258,097	2,465	1,258,097	2,465	0	0
O&M Cost ($1,000/yr)						
Fixed Costs	**Total**		**Total**		**Δ**	**% Δ**
Labor	24,051		24,051		0	0
Variable Operating Costs*	**Total**		**Total**			
Maintenance Materials	23,273		24,728		1,455	6
Water	1,434		1,524		90	6
Chemicals	1,969		2,092		123	6
Waste Disposal	2,502		2,659		157	6
Total Variable Costs	29,179		31,003		1,824	6
Total O&M Cost	53,230		55,054		1,824	3
Fuel Cost*	67,765		72,000		4,235	6
Discounted Cash Flow Results, levelized						
Capital Cost ($/kW-hr)	0.0616		0.0579			-6
Fixed O&M Cost ($/kW-hr)	0.0078		0.0073			-6
Variable O&M Cost ($/kW-hr)	0.0094		0.0094			0
Fuel Cost ($/kW-hr)	0.0228		0.0228			0
TS&M Cost ($/kW-hr)	0.0039		0.0039			0
Levelized COE ($/kW-hr)	0.1054		0.1014			-4

3.4. Increased Capacity Factor to 85 %

In this case, the process configuration and process performance remain the same as the previous case, but the capacity factor increases from 80 % to 85 %. The increased power production resulting from more time on-line reflects anticipated improvements in process reliability, availability, and maintainability (RAM) due to DOE-sponsored R&D in areas such as vessel refractories and improved sensors. In this analysis, it is assumed that these advancements add little additional capital or fixed O&M cost. The increased power production translates into additional revenue, which has a direct positive impact on the COE. Capital and O&M costs are compared in Table 3-8.

See Appendix A for NETL 's update to capital cost and COE.

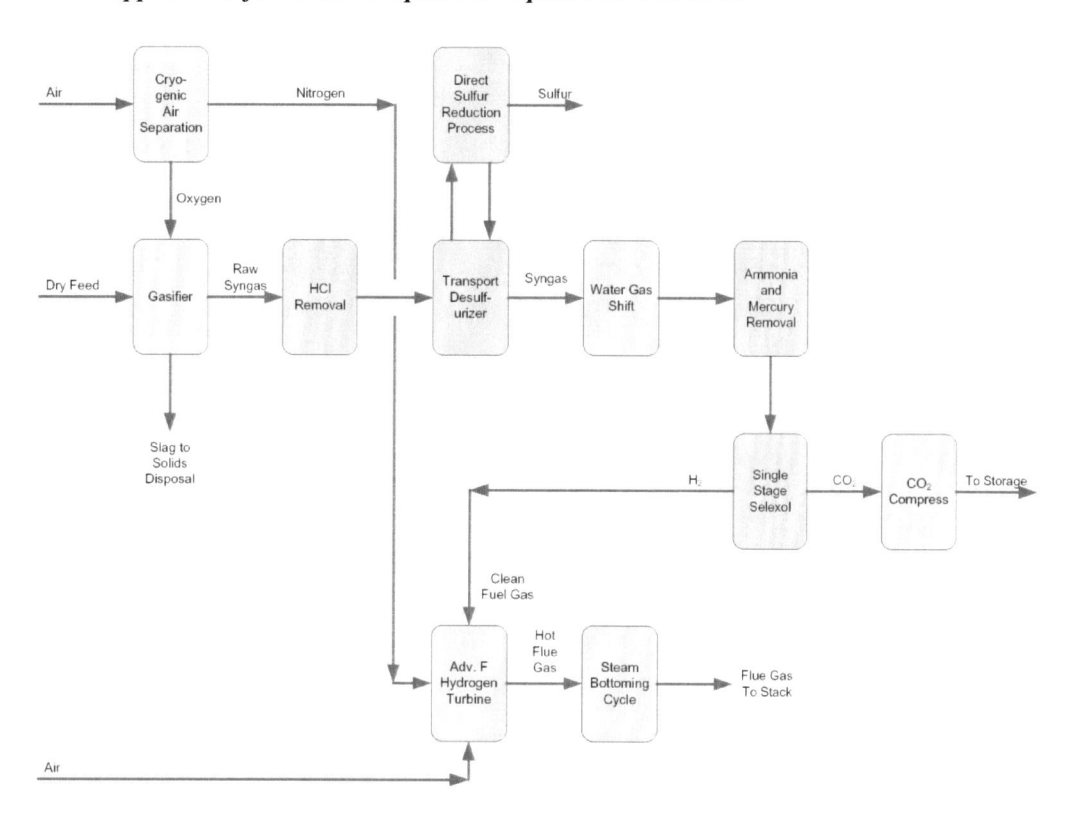

Figure 3-4. Warm Gas Cleanup With Selexol CO_2 Separation.

Capital cost is not affected by capacity factor, so the TPC is the same in both cases. The differences between cases lie in variable O&M costs and fuel cost, which increase by approximately 6 % as the result of increased annual hours of operation. However, the discounted cash flow spreads fixed costs over a greater amount of power production, more than compensating for these additional costs and resulting in an overall decrease in cost of electricity from $0.1054/kW-hr to $0.1014/kW-hr – a savings of about 4 % in cost of electricity resulting from increased capacity factor. On a percentage basis, this COE reduction is the same as the reduction for the corresponding non-capture analysis.

Table 3-9. Incremental Performance Improvement from Warm Gas Cleanup

	85 % Capacity Factor	WGCU + Selexol
Gas Turbine Power (MWe)	464	464
Fuel Gas Expander (MWe)	7	8
Steam Turbine Power (MWe)	228	258
Total Power Produced (MWe)	699	730
Auxiliary Power Use (MWe)	-189	-195
Net Power (MWe)	510	535
As-Received Coal Feed (lb/hr)	459,257	469,765
Net Heat Rate (Btu/kW-hr)	10,497	10,243
Net Plant Efficiency (HHV)	32.5 %	33.3 %

3.5. Warm Gas Cleanup with Selexol CO_2 Separation

In this case, the primary process improvement is that the cold gas ammonia scrub, mercury filter, Selexol H_2S removal, and Claus tail gas treatment processes are replaced with warm gas cleanup processes. A block flow diagram is presented in Figure 3-4. The warm gas transport desulfurization, direct sulfur reduction process (DSRP), novel ammonia removal, and mercury removal technologies are described in Volume 1 Section 3.6. When replacing the cold gas desulfurization section with warm gas desulfurization, the second-stage Selexol absorber is retained in order to separate CO_2 for sequestration.

Case Configuration: Coal Feed Pump, Cryogenic ASU, Warm Gas Cleanup, Single-Stage Selexol CO_2 Separation, Advanced "F" Hydrogen Turbine, 85 % Capacity Factor

The cold gas quench section is replaced with convective coolers and a chloride guard bed to remove HCl. This is followed by a transport desulfurizer with associated sorbent regenerator and DSRP.

Following desulfurization, two-stage shift, and warm gas ammonia and mercury removal, the H_2-rich syngas is quenched to remove water, and also to decrease temperature for entry to the Selexol absorber. The Selexol absorber produces low- and intermediate-pressure CO_2 streams that are directly compressed to sequestration pipeline pressure.

Table 3-9 compares process performance between cold gas cleanup and warm gas cleanup with Selexol CO_2 separation. Steam turbine power generation increases by 30 MW due to (1) elimination of the sour water stripper, (2) heat recovery during warm gas cleanup, and (3) greater heat recovery resulting from water gas shift.

The addition of (1) regeneration air compressor for warm gas cleanup, (2) increased N_2 compressor load for fuel diluent flow through the gas turbine, and (3) increased CO_2 compressor load due to increased flow of the CO_2 stream to sequestration are somewhat offset by reduced auxiliary load of the single-stage Selexol absorber, resulting in an auxiliary power increase by 6 MW.

With part of the desulfurized syngas used as reducing gas in the DSRP, slightly greater coal feed rate is needed for warm gas cleanup. The net impact of the higher auxiliary load and increased steam turbine power output is an increase of 25 MW, resulting in an increase in process efficiency from 32.5 % to 33.3 %.

Table 3-10. Warm Gas Cleanup With Selexol: Capital and O&M Cost Comparison

	85% Capacity Factor		WGCU + Selexol		Δ	
Capital Cost ($1 ,000)						
Plant Sections	**TPC**	**TPC $/kW**	**TPC**	**TPC $/kW**	**Δ TPC $/kW**	**% Δ**
1 Coal and Sorbent Handling	33,445	66	33,920	63	-3	-5
2 Coal and Sorbent Prep & Feed	55,442	109	56,073	105	-4	-4
3 Feedwater & Balance of Plant	34,231	67	34,503	64	-3	-4
4a Gasifier	247,284	485	257,684	482	-3	-1
4b Air Separation Unit	173,695	340	173,180	324	-16	-5
5a Gas Cleanup	226,119	443	240,416	449	6	1
5b CO_2 Removal & Compression	45,607	89	49,505	93	4	4
6 Gas Turbine	132,079	259	132,343	247	-12	-5
7 HRSG	53,439	105	53,848	101	-4	-4
8 Steam Cycle and Turbines	55,118	108	60,188	113	5	5
9 Cooling Water System	24,402	48	25,867	48	0	0
10 Waste Solids Handling System	39,732	78	40,291	76	-2	-3
11 Accessory Electric Plant	75,981	149	77,283	144	-5	-3
12 Instrumentation & Control	25,937	51	26,182	49	-2	-4
13 Site Preparation	18,958	37	19,050	36	-1	-3
14 Buildings and Structures	16,627	33	17,137	32	-1	-3
Total	1,258,097	2,465	1,297,471	2,425	-40	-2
O&M Cost ($1,000/yr)						
Fixed Costs	**Total**		**Total**		**Δ**	**% Δ**
Labor	24,051		24,051		0	0
Variable Operating Costs*	**Total**		**Total**			
Maintenance Materials	24,728		23,634		-1,094	-4
Water	1,524		1,567		43	3
Chemicals	2,092		6,076		3,984	190
Waste Disposal	2,659		2,720		61	2
Total Variable Costs	31,003		33,997		2,994	10
Total O&M Cost	55,054		58,049		2,995	5
Fuel Cost*	72,000		73,648		1,648	2
Discounted Cash Flow Results, levelized						
Capital Cost ($/kW-hr)	0.0579		0.0570			-2
Fixed O&M Cost ($/kW-hr)	0.0073		0.0070			-4
Variable O&M Cost ($/kW-hr)	0.0094		0.0099			5
Fuel Cost ($/kW-hr)	0.0228		0.0222			-3
TS&M Cost ($/kW-hr)	0.0039		0.0039			0
Levelized COE ($/kW-hr)	0.1014		0.1000			-1

*Includes 85 % Capacity Factor

Cost Analysis

See Appendix A for NETL 's update to capital cost and COE.

Capital and O&M costs are compared in Table 3-10. The gasifier cost in the WGCU with single- stage Selexol case increases due to increased coal feed rate and addition of the convective heat exchanger; however, due to the 25 MW increase in net power generation, the cost on a $/kW basis decreases slightly. Despite increase in coal feed rate, the ASU cost remains the same because of lower oxygen requirement with the elimination of the Claus plant; ASU cost on a $/kW basis decreases by $ 16/kW.

Although the cost of warm gas cleanup is significantly less than two-stage Selexol, the cost of gas cleanup increases by $14 MM ($6/kW) because of the incremental cost of the second-stage Selexol absorber.

The slight $4/kW increase in cost of CO_2 compression in the warm gas cleanup case is due to slightly greater coal feed rate and therefore increased CO_2 product, and also a slightly more dilute stream (and therefore increased flowrate) from single-stage Selexol than from the two- stage Selexol section.

Overall, TPC increases by $39 MM but because of increased net power output, capital cost decreases by $40/kW.

Variable O&M costs increase by 10 % in the warm gas cleanup case primarily due to the cost of ZnO sorbent used in the transport desulfurizer. Fuel cost increases slightly, due to the 2 % increase in coal feed rate to the process.

With only small variations in both capital and operating expenses, all terms resulting from the discounted cash flow calculation are very similar, with a net reduction in COE from $0.1014/kW-hr to $0.1000/kW-hr – a 1 % decrease.

3.6. Warm Gas Cleanup with Hydrogen Membrane

An innovative process technology, unique to the carbon capture configuration, is the hydrogen membrane which separates hydrogen from the warm syngas stream exiting the mercury and ammonia removal section. Hydrogen is removed at low partial pressure over two membrane stages; low partial pressure is achieved using N_2 sweep gas from the ASU. To purify for pipeline transport and sequestration, the CO_2-rich non-permeate is compressed to a liquid phase, and noncondensibles are separated and returned to the topping combustor. One benefit of the hydrogen membrane is that the CO_2 non-permeate is at high pressure, significantly reducing compressor load for sequestration.

Case Configuration: Coal Feed Pump, Cryogenic ASU, Warm Gas Cleanup, Hydrogen Membrane, Advanced "F" Hydrogen Turbine, 85 % Capacity Factor

Figure 3-5 shows a block flow diagram of this process configuration. Following transport desulfurization, the bulk of desulfurized syngas (already at 900 °F) is shifted in two stages. The high temperature shift operates at 650 °F, while the low temperature shift operates at 460 °F (a good temperature match for the novel ammonia and mercury removal section).

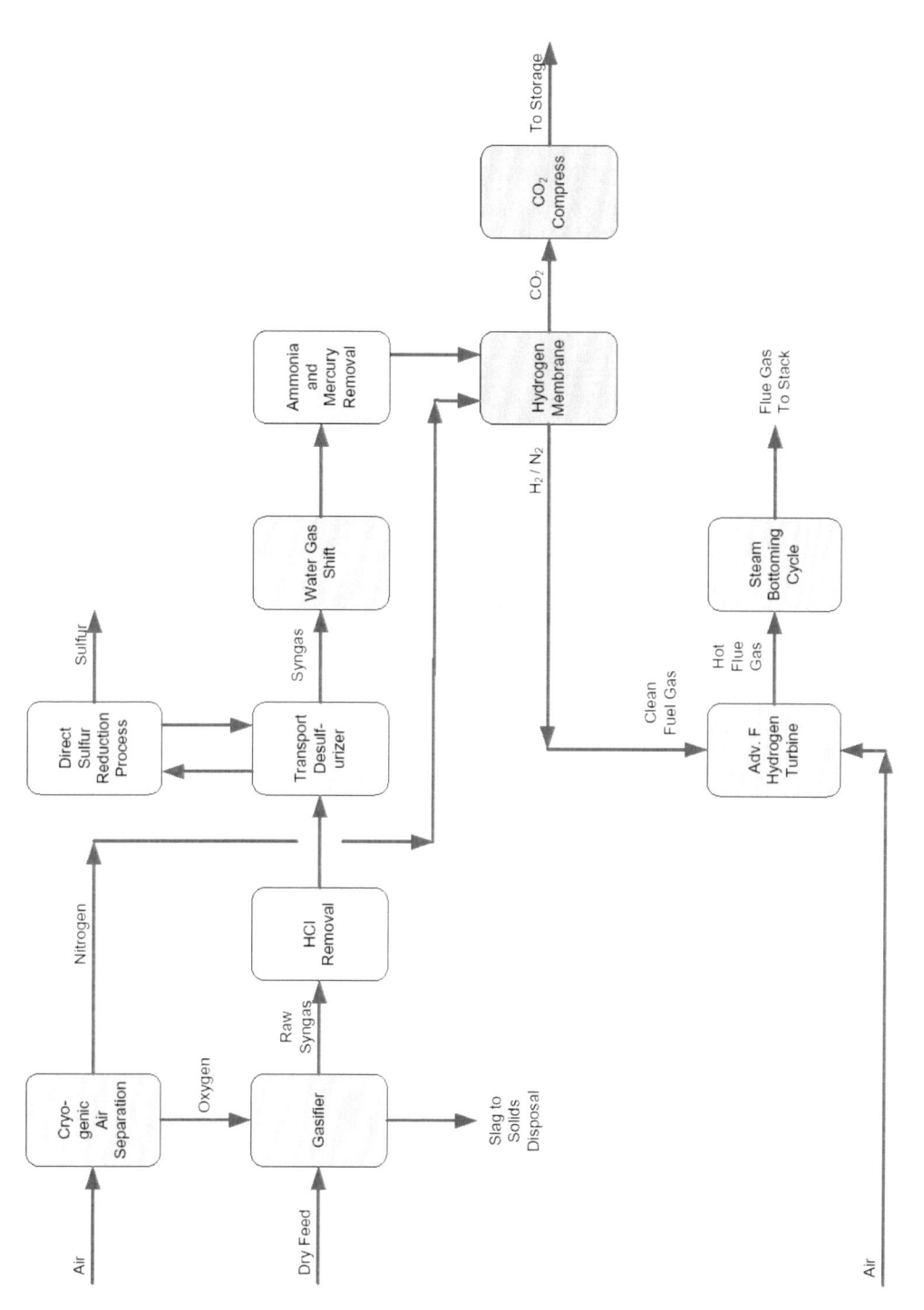

Figure 3-5. Warm Gas Cleanup With Hydrogen Membrane.

Sufficient steam must be added to convert CO to CO_2 in order to achieve 90 % carbon capture. The low temperature shift favors H_2 formation, which is why water gas shift in the H_2 membrane, operating at higher temperature (700 °F), is not desired.

Table 3-11. Incremental Performance Improvement from Hydrogen Membrane

	Warm Gas Cleanup + Selexol	Warm Gas Cleanup + H_2 Membrane
Gas Turbine Power (MWe)	464	464
Fuel Gas Expander (MWe)	8	NA
Steam Turbine Power (MWe)	258	267
Total Power Produced (MWe)	730	731
Auxiliary Power Use (MWe)	-195	-159
Net Power (MWe)	535	572
As-Received Coal Feed (lb/hr)	469,765	462,174
Net Heat Rate (Btu/kW-hr)	10,243	9,430
Net Plant Efficiency	33.3 %	36.2 %

Clean syngas from mercury and ammonia removal is reheated to the membrane operating temperature (700 °F is mid-range of anticipated operating temperatures), and then it enters a two- stage hydrogen membrane separator. Each membrane stage separates 68 % of the available H_2 for a total of 90 % recovery. The permeate pressure of each stage is set to the turbine fuel valve pressure. The fuel flowrate is set to achieve a turbine exit temperature of 1,050 °F. The net gas turbine power output is 232 MWe per unit.

The CO_2-rich non-permeate from the membrane is cooled for heat recovery, and moisture is removed. The CO_2 is compressed to 2,200 psig for transport to sequestration. During compression, the CO_2–rich stream, at slightly greater than 80 mole % purity, is condensed in order to recover impurities (primarily N_2, CO, and H_2) which are returned to the topping combustor.

All available process heat is collected for steam generation in the bottoming cycle. Superheated steam is expanded through three turbines, with reheat after the high pressure turbine. The bottoming cycle also provides heat for shift steam generation.

Table 3-11 summarizes the overall performance for two process trains. Heat recovery increases in the hydrogen membrane case as the result of eliminating the Selexol reboiler duty, thereby increasing steam turbine power by 9 MW.

Despite losing 8 MW from the fuel gas expander, the 9 MW increase in steam turbine power generation and the 36 MW decrease in auxiliary power results in a 37 MW increase in net power generation. The primary contributions to the decrease in auxiliary power are a 23 MW (60 %) reduction in CO_2 compression (because of high CO_2 delivery pressure from the hydrogen membrane) and elimination of Selexol auxiliaries for 13 MW.

With a slight decrease in coal feed rate, the net result is a plant efficiency increase by 2.9 percentage points from 33.3 % to 36.2 %.

Cost Analysis

See Appendix A for NETL 's update to capital cost and COE.

Capital and O&M costs are compared in Table 3-12. Total plant cost for coal handling, coal feed, gasifier, ASU, and general plant (cooling water system, waste handling, site preparation, and buildings) accounts are very similar because the coal flowrates in both cases are nearly the same; TPC decreases by about 7 % on a $/kW basis for these accounts in the hydrogen membrane case because of greater net power production.

Gas cleanup cost decreases significantly due to replacing the gas quench, second-stage Selexol absorber, and fuel reheat equipment with the less-expensive H_2 membrane; the net reduction in gas cleanup cost is $92 MM, and the TPC reduction on a $/kW basis is $ 189/kW. The bare erected cost of the hydrogen membrane is based on a technology development target cost of $450 per square foot of membrane surface area, and with a service life of 5 years.

Table 3-12. WGCU/H_2 Membrane: Capital and O&M Cost Comparison

	WGCU + Selexol		WGCU + H_2 Membrane		Δ	
Capital Cost ($1 ,000)						
Plant Sections	**TPC**	**TPC $/kW**	**TPC**	**TPC $/kW**	**Δ TPC $/kW**	**% Δ**
1 Coal and Sorbent Handling	33,920	63	33,576	59	-4	-6
2 Coal and Sorbent Prep & Feed	56,073	105	55,457	97	-8	-8
3 Feedwater & Balance of Plant	34,503	64	34,308	60	-4	-6
4a Gasifier	257,684	482	255,212	446	-36	-7
4b Air Separation Unit	173,180	324	178,584	312	-12	-4
5a Gas Cleanup	240,416	449	148,432	260	-189	-42
5b CO_2 Removal & Compression	49,505	93	25,392	44	-49	-53
6 Gas Turbine	132,343	247	124,363	218	-29	-12
7 HRSG	53,848	101	53,803	94	-7	-7
8 Steam Cycle and Turbines	60,188	113	61,669	108	-5	-4
9 Cooling Water System	25,867	48	26,288	46	-2	-4
10 Waste Solids Handling System	40,291	76	39,888	70	-6	-8
11 Accessory Electric Plant	77,283	144	73,141	128	-16	-11
12 Instrumentation & Control	26,182	49	24,716	43	-6	-12
13 Site Preparation	19,050	36	18,723	33	-3	-8
14 Buildings and Structures	17,137	32	17,111	30	-2	-6
Total	1,297,471	2,425	1,170,662	2,047	-378	-16

Table 3-12. (Continued)

	WGCU + Selexol	WGCU + H₂ Membrane	Δ	
O&M Cost ($1,000/yr)				
Fixed Costs	**Total**	**Total**	**Δ**	**% Δ**
Labor	24,051	22,548	-1,503	-6
Variable Operating Costs*	**Total**	**Total**		
Maintenance Materials	23,634	23,656	22	0
Water	1,567	1,449	-118	-8
Chemicals	6,076	5,688	-388	-6
Membrane Replacement	NA	945	945	∞
Waste Disposal	2,720	2,675	-45	-2
Total Variable Costs	33,997	34,414	417	1
Total O&M Cost	**58,049**	**56,963**	**-1,086**	**-2**
Fuel Cost*	**73,648**	**72,458**	**-1,190**	**-2**
Discounted Cash Flow Results, levelized				
Capital Cost ($/kW-hr)	0.0570	0.0481		-16
Fixed O&M Cost ($/kW-hr)	0.0070	0.0061		-13
Variable O&M Cost ($/kW-hr)	0.0099	0.0094		-5
Fuel Cost ($/kW-hr)	0.0222	0.0205		-8
TS&M Cost ($/kW-hr)	0.0039	0.0039		0
Levelized COE ($/kW-hr)	0.1000	0.0880		-12

*Includes 85 % Capacity Factor

CO_2 compression cost decreases by $49/kW in the hydrogen membrane case because of decreased CO_2 compressor load. The high pressure of the non-permeate stream exiting the membrane allows expansion to provide auto-refrigeration to condense and separate CO_2, and the pressure of the expanded stream is still greater than recovery pressure from Selexol.

The cost of the gas turbine account decreases by $8 MM due to elimination of the syngas expander, resulting in a further reduction of $2 9/kW in TPC. Overall, the TPC of the hydrogen membrane case decreases by $378/kW. O&M costs decrease slightly as the O&M cost is roughly a function of TPC. Fuel cost decreases by 2 % resulting from improved process efficiency in the hydrogen membrane case.

The $92 MM reduction in TPC of warm gas cleanup with the H_2 membrane compared to the cost of warm gas cleanup with second-stage Selexol cold gas cleanup process represents the primary cost advantage of this case. A secondary cost incentive is the increase in net power produced by the hydrogen membrane case, which further reduces the TPC on a $/kW basis. Compared to the Selexol process, CO_2 separation via the hydrogen membrane is projected to reduce the levelized COE from $0.1000/kW-hr to $0.0880/kW-hr – a decrease of 12 %.

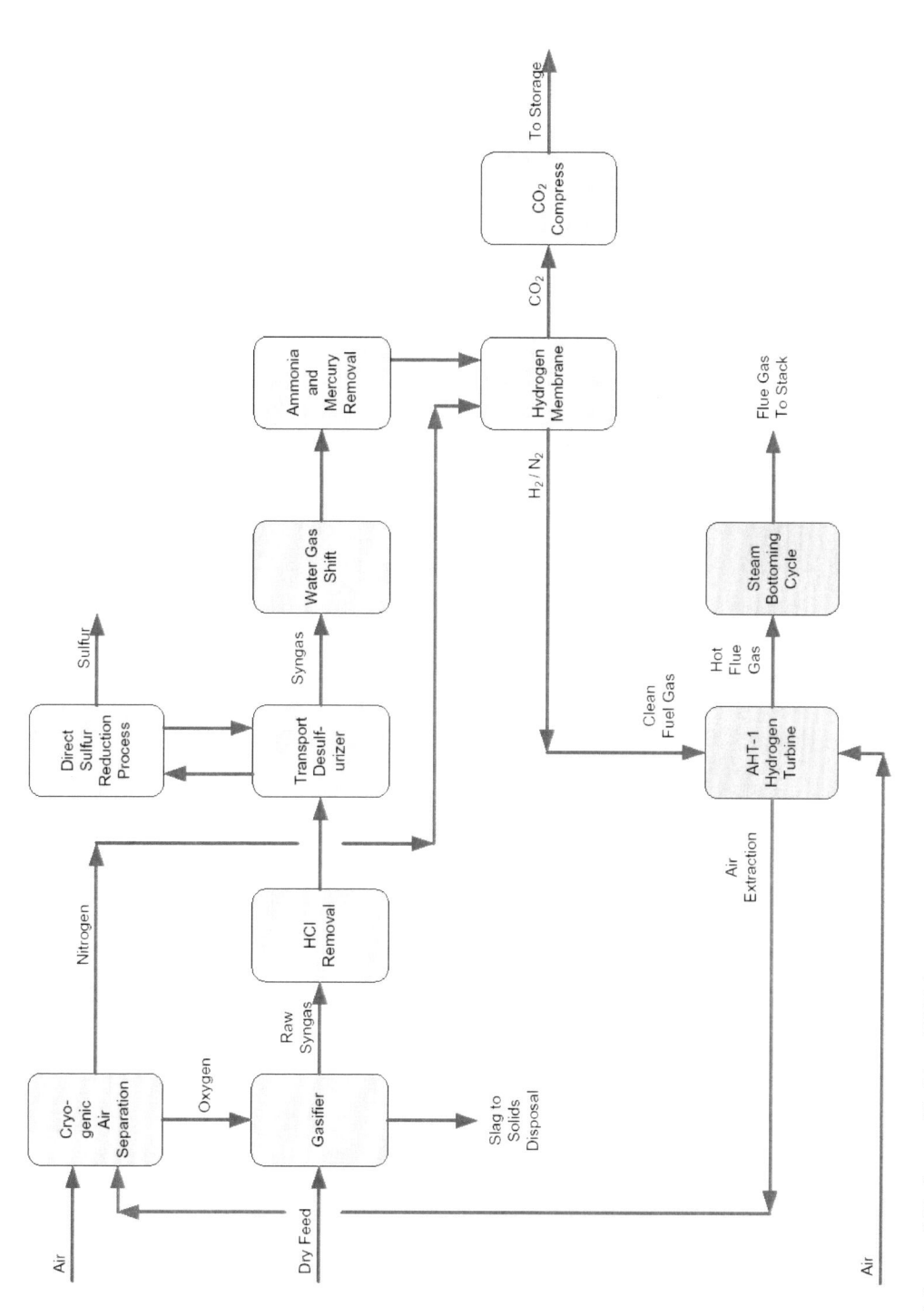

Figure 3-6. Advanced Hydrogen Turbine AHT-1.

3.7. Advanced Hydrogen Turbine, First Generation (AHT-1)

DOE sponsors R&D to develop advanced turbine technology with improved performance efficiency. For the purposes of this analysis, this advanced hydrogen turbine is named AHT-1. Performance improvement is expected primarily from higher turbine inlet temperature, which will improve efficiency of the turbine over exiting state-of-the-art. A block flow diagram of an advanced turbine case is presented in Figure 3-6. In addition to modified turbine performance parameters, steam cycle superheat and reheat temperatures increase to 1,050 °F resulting from increased turbine exit temperature, and air integration becomes possible.

Case Configuration: Coal Feed Pump, Cryogenic ASU, Warm Gas Cleanup, Hydrogen Membrane, AHT-1 Turbine, 85 % Capacity Factor

Table 3-13 demonstrates improved overall process performance when the advanced "F" hydrogen turbine is replaced with a somewhat larger and more advanced A HT-1 turbine.

Gas turbine power increases by 36 MW due to the improved A HT-1 turbine. The higher pressure ratio and slightly greater throughput contribute to improved turbine performance. The turbine exit temperature limitation of 1,050 °F is lifted in the AHT-1 turbine due to expectations that R& D will provide improved materials to withstand high flue gas moisture content.

The 40 MW increase in steam turbine power results somewhat from increased coal feed rate (and associated process and HRSG heat recovery), but more importantly from increased steam superheat and reheat temperature to 1,050 °F which improves the heat rate (Carnot efficiency) of the bottoming cycle.

Auxiliary power use decreases by 11 MW due to air integration, which decreases the parasitic load on the ASU main air compressor.

Increased steam turbine power and reduced auxiliary power, together with a significant increase in gas turbine power, are responsible for the increased process efficiency from 36.2 % to 38.0 % – an increase of 1.8 percentage points. Because of the H_2-rich fuel in the carbon capture cases, operating constraints limit the performance of the advanced "F" turbine (introduced previously), specifically; (1) gas turbine and steam cycle performance are lower than in a non-capture scenario because of turbine exhaust temperature limit, and (2) due to the reduced volume of H_2- rich gas relative to syngas, no air integration is possible, which impacts ASU auxiliary load. These constraints are removed with advancement to the A HT-1 and, since those constraints never applied to the non-capture advanced turbine case, the impact of the AHT-1 advancement is greater in the carbon capture case (1.8 percentage point improvement) than in the non-capture case (1.0 percentage point improvement).

Cost Analysis

See Appendix A for NETL 's update to capital cost and COE.

Capital and O&M costs are compared with results from the previous case in Table 3-14. Total plant cost for all sections increases due to the increased plant size. Because the AHT-1 produces more power, TPC decreases on a $/kW basis for all cost accounts.

Table 3-13. Incremental Performance Improvement from the AHT-1 Turbine

	WGCU+H$_2$ Membrane	AHT-1 Turbine
Gas Turbine Power (MWe)	464	500
Steam Turbine Power (MWe)	267	307
Total Power Produced (MWe)	731	807
Auxiliary Power Use (MWe)	-159	-148
Net Power (MWe)	572	659
As-Received Coal Feed (lb/hr)	462,174	506,903
Net Heat Rate (Btu/kW-hr)	9,430	8,976
Net Plant Efficiency (HHV)	36.2 %	38.0 %

Table 3-14. AHT-1 Turbine: Capital and O&M Cost Comparison

	WGCU + H$_2$ Membrane		AHT-1 Turbine		Δ	
Capital Cost ($1,000)						
Plant Sections	TPC	TPC $/kW	TPC	TPC $/kW	Δ TPC $/kW	% Δ
1 Coal and Sorbent Handling	33,576	59	35,559	54	-5	-8
2 Coal and Sorbent Prep & Feed	55,457	97	59,024	90	-7	-7
3 Feedwater & Balance of Plant	34,308	60	35,442	54	-6	-10
4a Gasifier	255,212	446	271,147	412	-34	-8
4b Air Separation Unit	178,584	312	180,416	274	-38	-12
5a Gas Cleanup	148,432	260	159,141	242	-18	-7
5b CO$_2$ Removal & Compression	25,392	44	27,860	42	-2	-5
6 Gas Turbine	124,363	218	125,785	191	-27	-12
7 HRSG	53,803	94	55,802	85	-9	-10
8 Steam Cycle and Turbines	61,669	108	68,004	103	-5	-5
9 Cooling Water System	26,288	46	27,662	42	-4	-9
10 Waste Solids Handling System	39,888	70	42,224	64	-6	-9
11 Accessory Electric Plant	73,141	128	73,134	111	-17	-13
12 Instrumentation & Control	24,716	43	24,207	37	-6	-14
13 Site Preparation	18,723	33	18,795	29	-4	-12
14 Buildings and Structures	17,111	30	17,654	27	-3	-10
Total	1,170,662	2,047	1,221,858	1,855	-192	-9
O&M Cost ($1,000/yr)						
Fixed Costs	**Total**		**Total**		Δ	% Δ
Labor	22,548		24,051		1,503	7
Variable Operating Costs*	**Total**		**Total**			
Maintenance Materials	23,656		25,370		1,714	7
Water	1,449		1,508		59	4
Chemicals	5,688		6,245		557	10
Membrane Replacement	945		1,041		96	10

Table 3-14. (Continued)

Variable Operating Costs*	WGCU + H₂ Membrane Total	AHT-1 Turbine Total	Δ	
Waste Disposal	2,675	2,935	260	10
Total Variable Costs	34,414	37,098	2,684	8
Total O&M Cost	56,963	61,150	4,187	7
Fuel Cost*	72,458	79,470	7,012	10
Discounted Cash Flow Results, levelized				
Capital Cost ($/kW-hr)	0.0481	0.0436		-9
Fixed O&M Cost ($/kW-hr)	0.0061	0.0057		-7
Variable O&M Cost ($/kW-hr)	0.0094	0.0087		-7
Fuel Cost ($/kW-hr)	0.0205	0.0195		-5
TS&M Cost ($/kW-hr)	0.0039	0.0039		0
Levelized COE ($/kW-hr)	0.0880	0.08 14		-8

*Includes 85% Capacity Factor

The cost of the turbine is scaled to the turbine power rating; the increase in power rating of the 232 MW advanced "F" turbine to the 250 MW AHT-1 turbine increases turbine cost by $1,422 K. No cost premium is assumed for higher temperature operation. After accounting for the net power increase in the AHT-1 case, turbine cost decreases by $27/kW.

The TPC decreases by $192/kW as a result of the AHT-1 turbine. This is significantly greater than the $72/kW cost reduction in the non-capture scenario. Contributions of increased steam superheat/reheat temperature and air integration when transitioning from the advanced "F" turbine to the AHT-1 turbine result in an 87 MW increase in net plant power output which, when divided into the TPC, decreases TPC on a $/kW basis more than in the non-capture scenario. The cost reduction is not so much the result of the turbine cost, but the additional power generated by the plant as a consequence of the improved turbine.

The increased O&M and fuel costs reflect larger plant size and increased coal throughput. The net reduction in COE from $0.0880/kW-hr to $0.0814/kW-hr represents a 6.6 mills/kW-hr decrease in COE resulting from the AHT-1 turbine. The non-capture scenario, by comparison, results in a 2.7 mills/kW-hr decrease in COE.

3.8. Ion Transport Membrane

In this case, an ITM replaces the cryogenic ASU for oxygen production. Oxygen diffuses through a ceramic wall in the ITM based on partial pressure driving force, and leaves the nitrogen-rich non-permeate as secondary product. The non-permeate remains at high pressure, which is essentially the feed pressure to the ITM, while the oxygen permeate stream is produced at as low a pressure as possible in order to maximize partial pressure driving force for the separation and to reduce oxygen concentration in the non-permeate to as low as 2 mole %. The high pressure of the non-permeate stream is one of the advantages of the ITM; it eliminates the need for the N2 compressor – reducing auxiliary power consumption, but that is partially offset by the increased power consumption of the ITM boost and oxygen compressors. The primary advantage of the ITM, however, is the reduced capital cost of air separation relative to a cryogenic ASU.

Case Configuration: Coal Feed Pump, Ion Transport Membrane, Warm Gas Cleanup, Hydrogen Membrane, AHT-1 Turbine, 85 % Capacity Factor

A block flow diagram of this process is shown in Figure 3-7.

The fraction of air integration is varied in order to meet the turbine power rating of 250 MW per unit. Coal feed rate (and therefore fuel flow) is adjusted to satisfy the turbine inlet temperature of 2,550 °F. Table 3-15 below compares overall process performance improvement due to air separation using the ITM.

Steam turbine power increases by 40 MW in the ITM case due to increased coal feed rate (and therefore heat recovery throughout the process) and also heat recovery from hot sweep gas from the ITM to the hydrogen membrane (as opposed to heating cold sweep gas from the cryogenic ASU in the previous case).

Although elimination of the nitrogen compressor in the ITM case decreases auxiliary load, it is counterbalanced by the ITM boost compressor and the oxygen compressor loads. The net auxiliary power increases by 8 MW.

Table 3-15. Incremental Performance Improvement from the ITM

	AHT-1 Turbine	ITM
Gas Turbine Power (MWe)	500	500
Steam Turbine Power (MWe)	307	347
Total Power Produced (MWe)	807	847
Auxiliary Power Use (MWe)	-148	-156
Net Power (MWe)	659	691
As-Received Coal Feed (lb/hr)	506,903	527,717
Net Heat Rate (Btu/kW-hr)	8,976	8,908
Net Plant Efficiency (HHV)	38.0 %	38.3 %

The additional 32 MW net power generated in the ITM case is accompanied by increased coal feed required to (1) provide fuel to heat the ITM, and (2) to produce 112 to consume residual oxygen in the sweep gas before it is introduced to the hydrogen membrane. The ITM process results in a 0.3 percentage point improvement in net plant efficiency for the carbon capture scenario.

In the non-capture scenario, process efficiency increases by 0.65 percentage points, and coal feed rate remains essentially unchanged. Of the fuel gas generated in the non-capture ITM case, 10 % of it is used to heat the ITM; in the carbon capture ITM case, only 1 % of the H_2 fuel stream is used to heat the ITM. A recuperator is responsible for reducing the amount of fuel needed to heat the ITM in the carbon capture case. Per discussion with the ITM technology developer, the recuperator would be appropriate for the carbon capture case but not for the non-capture case.

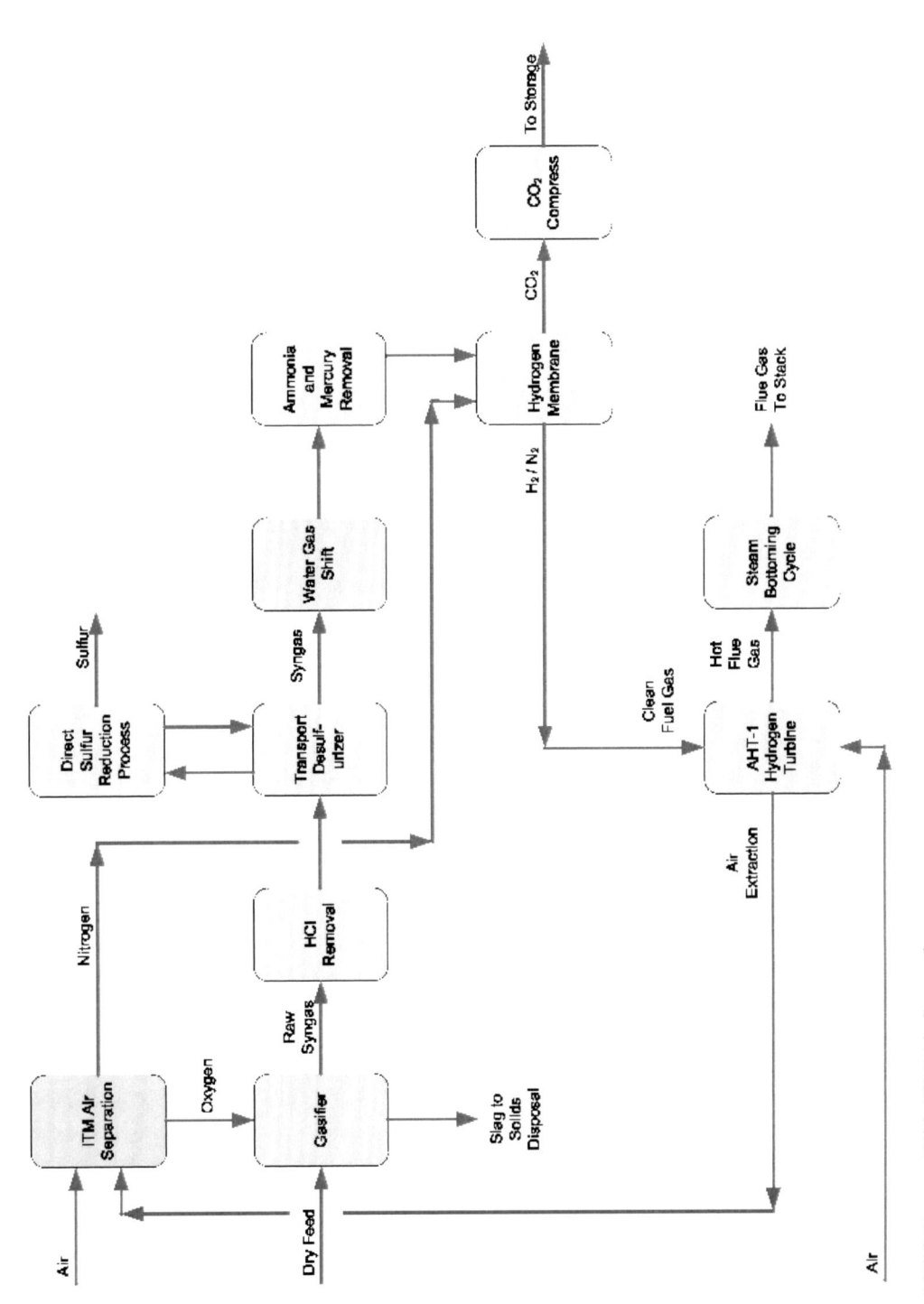

Figure 3-7. IGCC Process With ITM Air Separation.

Table 3-16. ITM: Capital and O&M Cost Summary

Plant Sections	AHT-1 Turbine		ITM		Δ	
Capital Cost ($1 ,000)						
Plant Sections	TPC	TPC $/kW	TPC	TPC $/kW	Δ TPC $/kW	% Δ
1 Coal and Sorbent Handling	35,559	54	36,457	53	-1	-2
2 Coal and Sorbent Prep & Feed	59,024	90	60,651	88	-2	-2
3 Feedwater & Balance of Plant	35,442	54	35,957	52	-2	-4
4a Gasifier	271,147	412	277,047	401	-11	-3
4b Air Separation Unit	180,416	274	120,312	174	-100	-36
5a Gas Cleanup	159,141	242	167,120	242	0	0
5b CO_2 Removal & Compression	27,860	42	28,687	42	0	0
6 Gas Turbine	125,785	191	125,785	182	-9	-5
7 HRSG	55,802	85	55,904	81	-4	-5
8 Steam Cycle and Turbines	68,004	103	74,327	108	5	5
9 Cooling Water System	27,662	42	29,355	42	0	0
10 Waste Solids Handling System	42,224	64	43,285	63	-1	-2
11 Accessory Electric Plant	73,134	111	74,921	108	-3	-3
12 Instrumentation & Control	24,207	37	24,577	36	-1	-3
13 Site Preparation	18,795	29	18,954	27	-2	-7
14 Buildings and Structures	17,654	27	18,283	26	-1	-4
Total	1,221,858	1,855	1,191,624	1,724	-131	-7
O&M Cost ($1,000/yr)						
Fixed Costs	**Total**		**Total**		Δ	% Δ
Labor	24,051		22,548		-1,503	-6
Variable Operating Costs*	**Total**		**Total**			
Maintenance Materials	25,370		26,284		914	4
Water	1,508		1,470		-38	-3
Chemicals	6,245		6,535		290	5
Membrane Replacement	1,041		987		-54	-5
Waste Disposal	2,935		3,055		120	4
Total Variable Costs	37,098		38,331		1,233	3
Total O&M Cost	61,150		60,880		-270	0
Fuel Cost*	79,470		82,733		3,263	4
Discounted Cash Flow Results, levelized						
Capital Cost ($/kW-hr)	0.0436		0.0405			-7
Fixed O&M Cost ($/kW-hr)	0.0057		0.005 1			-11
Variable O&M Cost ($/kW-hr)	0.0087		0.0086			-1
Fuel Cost ($/kW-hr)	0.0195		0.0193			-1
TS&M Cost ($/kW-hr)	0.0039		0.0039			0
Levelized COE ($/kW-hr)	0.08 14		0.0774			-5

*Includes 85 % Capacity Factor

Cost Analysis

See Appendix A for NETL's update to capital cost and COE.

The ITM cost includes main air compressor, ITM boost compressor, recuperator, two membrane stages, air heater, oxygen coolers, oxygen compressors, fluff gas cooler, and fluff gas compressor. The capital cost of the ITM section is assumed to be the target development cost of 67 % that of a comparable cryogenic ASU plant.

Comparing capital costs in Table 3-16, total plant cost for coal handling, coal feed, gasifier, gas cleanup, CO_2 compression, and general plant systems (feedwater, cooling water system, waste handling, site preparation, and buildings) are similar because of similar coal feed rates. Because the ITM case produces 32 MW more power than the cryogenic case, TPC decreases slightly on a $/kW basis for these cost accounts.

The cost of the ASU decreases significantly because the ITM costs 1/3 less than a cryogenic ASU. Coupled with the increased power production, the cost reduction by $100/kW for the ASU is the single greatest contribution to the overall plant TPC reduction.

Gas turbine cost is unchanged. Considering the additional power generation in the ITM case, however, the gas turbine cost decreases by $9/kW. Steam turbine cost increases by $5/kW for the ITM case due to greater heat recovery and steam turbine power generation.

Overall, the $131/kW reduction in TPC is primarily due to capital cost savings in the ASU. The second most important factor in the cost reduction is the 32 MW increase in power generated by the ITM case.

Table 3-17. Incremental Performance Improvement from AHT-2 Turbine

	ITM	AHT-2 Turbine
Gas Turbine Power (MWe)	500	370
Steam Turbine Power (MWe)	347	232
Total Power Produced (MWe)	847	602
Auxiliary Power Use (MWe)	-156	-100
Net Power (MWe)	691	502
As-Received Coal Feed (lb/hr)	527,717	366,990
Net Heat Rate (Btu/kW-hr)	8,908	8,524
Net Plant Efficiency (HHV)	38.3 %	40.0 %

O&M costs remain nearly the same, and fuel cost increases by 4 % due to increased coal feed rate. The reduction in COE from $0.0814/kW-hr to $0.0774/kW-hr, therefore, is due primarily to the decrease in capital cost of the ASU and increased net power production as the result of the ITM.

In the non-capture cases, TPC decreases by $82/kW as the result of the ITM. The cost reduction is somewhat greater in the carbon capture scenario, with a $131 /kW decrease; the primary factor for larger decrease is the larger ASU required because of increased coal flow in carbon capture scenarios, and therefore greater potential for cost savings. The cost savings in ASU alone is $64/kW in the non-capture scenario versus $ 100/kW in the carbon capture scenario.

The capital cost reduction due to the ITM is reflected in greater reduction in COE in the carbon capture scenario (by 4.0 mills/kW-hr) than in the non-capture scenario (by 2.6 mills/kW-hr).

3.9. Next Generation Advanced Hydrogen Turbine (AHT-2)

DOE sponsors research to develop a turbine with even further improved performance over that of the AHT- 1. This is projected to be accomplished with higher firing temperature, increased power rating, and improved stage efficiencies. The pseudonym AHT-2 is used to refer to this advanced hydrogen turbine.

Case Configuration: Dry Feed Gasifier, ITM, Warm Gas Cleanup, Hydrogen Membrane, AHT-2 Turbine, 85 % Capacity Factor

The process block flow diagram of this IGCC process with the AHT-2 hydrogen turbine is identical to Figure 3-7 above. A single train produces a net 502 MW of power. Overall efficiency is 40.0 % (HHV basis). Carbon utilization is 99.5 % and the capacity factor is 85 %. Performance resulting from the AHT-2 turbine is compared to the AHT-1 in Table 3-17.

The overall decrease in net power generation is due to reducing the plant from two trains of AHT- 1 turbines to a single train of AHT-2 turbine in order to maintain the nominal plant output of 600 MW. The decrease in coal feed rate results in less steam turbine power generation and less auxiliary power from a smaller plant. Net plant efficiency improves by 1.7 percentage points as the result of higher pressure ratio and improved engine efficiency of the AHT-2.

In the non-capture scenario, introduction of the next-generation advanced syngas turbine increases process efficiency by 2.0 percentage points above that of the first generation advanced turbine. The efficiency improvement is dampened in the carbon capture scenario because of (1) increased coal feed rate per MW of gas turbine power in carbon capture versus non-capture cases, and (2) increased auxiliary power for oxygen production (resulting from increased coal feed) and CO_2 compression.

Cost Analysis

See Appendix A for NETL 's update to capital cost and COE.

Capital and O&M costs are compared in Table 3-18. The TPC in all accounts decreases because of reduced coal flowrate and decreased plant equipment size, and therefore cost. The number of process trains (consisting of gasifier, ASU, gas cleanup, CO_2 compression, and gas turbine) decreases from two to one. In each of these process sections, TPC on a $/kW basis decreases because of economy of scale for a single large train. All other process section accounts increase on a $/kW basis because of the decrease in net power production; this introduces a reverse economy of scale for those other process sections.

Table 3-18. AHT-2 Turbine (Single-Train): Capital and O& M Cost Summary

	ITM		AHT-2 Turbine (Single Train)		Δ	
Capital Cost ($1 ,000)						
Plant Sections	**TPC**	**TPC $/kW**	**TPC**	**TPC $/kW**	**Δ TPC $/kW**	**% Δ**
1 Coal and Sorbent Handling	36,457	53	29,100	58	5	9
2 Coal and Sorbent Prep & Feed	60,651	88	47,465	95	7	8
3 Feedwater & Balance of Plant	35,957	52	31,740	63	11	21
4a Gasifier	277,047	401	173,608	346	-55	-14
4b Air Separation Unit	120,312	174	77,413	154	-20	-11
5a Gas Cleanup	167,120	242	109,378	218	-24	-10
5b CO_2 Removal & Compression	28,687	42	19,957	40	-2	-5
6 Gas Turbine	125,785	182	83,208	166	-16	-9
7 HRSG	55,904	81	41,401	82	1	1
8 Steam Cycle and Turbines	74,327	108	55,760	111	3	3
9 Cooling Water System	29,355	42	24,243	48	6	14
10 Waste Solids Handling System	43,285	63	34,607	69	6	10
11 Accessory Electric Plant	74,921	108	62,181	124	16	15
12 Instrumentation & Control	24,577	36	21,672	43	7	19
13 Site Preparation	18,954	27	17,652	35	8	30
14 Buildings and Structures	18,283	26	16,184	32	6	23
Total	1,191,624	1,724	845,569	1,683	-41	-2
O&M Cost ($1 ,000/yr)						
Fixed Costs	**Total**		**Total**		**Δ**	**% Δ**
Labor	22,548		16,535		-6,013	-27
Variable Operating Costs*	**Total**		**Total**			
Maintenance Materials	26,284		20,751		-5,533	-21
Water	1,470		1,110		-360	-24
Chemicals	6,535		4,565		-1,970	-30
Membrane Replacement	987		692		-295	-30
Waste Disposal	3,055		2,125		-930	-30
Total Variable Costs	38,331		29,242		-9,089	-24
Total O&M Cost	60,880		45,777		-15,103	-25
Fuel Cost*	82,733		57,535		-25,198	-30
Discounted Cash Flow Results, levelized						
Capital Cost ($/kW-hr)	0.0405		0.03 96			-2
Fixed O&M Cost ($/kW-hr)	0.005 1		0.005 1			0
Variable O&M Cost ($/kW-hr)	0.0086		0.0090			5
Fuel Cost ($/kW-hr)	0.0193		0.0185			-4
TS&M Cost ($/kW-hr)	0.003 9		0.0039			0
Levelized COE ($/kW-hr)	0.0774		0.0761			-2

*Includes 85 % Capacity Factor

Table 3-19. AHT-2 Turbine (Two-Train): Capital and O&M Cost Summary

	ITM		AHT-2 Turbine (Two Trains)		Δ	
Capital Cost ($1,000)						
Plant Sections	**TPC**	**TPC $/kW**	**TPC**	**TPC $/kW**	**Δ TPC $/kW**	**% Δ**
1 Coal and Sorbent Handling	36,457	53	44,741	45	-8	-15
2 Coal and Sorbent Prep & Feed	60,651	88	75,780	75	-13	-15
3 Feedwater & Balance of Plant	35,957	52	40,708	41	-11	-21
4a Gasifier	277,047	401	344,033	342	-59	-15
4b Air Separation Unit	120,312	174	151,449	151	-23	-13
5a Gas Cleanup	167,120	242	213,905	213	-29	-12
5b CO_2 Removal & Compression	28,687	42	39,914	40	-2	-5
6 Gas Turbine	125,785	182	166,417	166	-16	-9
7 HRSG	55,904	81	68,483	68	-13	-16
8 Steam Cycle and Turbines	74,327	108	91,706	91	-17	-16
9 Cooling Water System	29,355	42	33,800	34	-8	-19
10 Waste Solids Handling System	43,285	63	53,045	53	-10	-16
11 Accessory Electric Plant	74,921	108	86,127	86	-22	-20
12 Instrumentation & Control	24,577	36	26,385	26	-10	-28
13 Site Preparation	18,954	27	20,075	20	-7	-26
14 Buildings and Structures	18,283	26	20,049	20	-6	-23
Total	1,191,624	1,724	1,476,615	1,470	-254	-15
O&M Cost ($1,000/yr)						
Fixed Costs	**Total**		**Total**		**Δ**	**% Δ**
Labor	22,548		28,561		6,013	27
Variable Operating Costs*	**Total**		**Total**			
Maintenance Materials	26,284		34,375		8,091	31
Water	1,470		1,768		298	20
Chemicals	6,535		9,124		2,589	40
Membrane Replacement	987		1,383		396	40
Waste Disposal	3,055		4,249		1,194	39
Total Variable Costs	38,331		50,900		12,569	33
Total O&M Cost	60,880		79,461		18,581	31
Fuel Cost*	82,733		115,070		32,337	39
Discounted Cash Flow Results, levelized						
Capital Cost ($/kW-hr)	0.0405		0.0345			-15
Fixed O&M Cost ($/kW-hr)	0.005 1		0.0044			-14
Variable O&M Cost ($/kW-hr)	0.0086		0.0079			-8
Fuel Cost ($/kW-hr)	0.0193		0.0185			-4
TS&M Cost ($/kW-hr)	0.0039		0.0039			0
Levelized COE ($/kW-hr)	0.0774		0.0692			-11

*Includes 85 % Capacity Factor

Overall, the total plant cost decreases by $346 MM going to a single train of the AHT-2 turbine. On a $/kW basis, the carbon capture plant with the AHT-2 turbine decreases by $41/kW or 2 %. In the non-capture cases, by comparison, TPC decreases by $319 MM and by $1 5/kW on a $/kW basis. The effect of the A HT-2 turbine on TPC is nearly the same in both capture and non- capture scenarios.

O&M cost reductions going from the two-train AHT-1 case to the single train A HT-2 case are also very similar between both non-capture and the capture scenarios.

The COE reduction from $0.0774/kW-hr to $0.0761/kW-hr in the carbon capture scenario (by 2 %) is similar to the 1 % decrease in COE in the non-capture scenario.

Two-Train Configuration

The discussion above features a single-train AHT-2 configuration that is constrained by the nominal plant size of 600 MW, which is the basis for this study. That process encounters a reverse economy of scale when the net plant power output is reduced to only 502 MW. If the process were allowed to maintain two power trains, with a net plant output of 1,004 MW, the process economics presented in Table 3-19 benefit from economy of scale compared to the previous case with the AHT-1 turbine.

The TPC in all accounts increases because of increased net power production, which corresponds to increased coal flowrate and increased plant equipment size, and therefore cost. On a $/kW basis, however, TPC decreases in all capital cost accounts. The bottom-line TPC decreases from $1 ,724/kW for the A HT-1 plant to $1 ,470/kW for the A HT-2 plant – a decrease of 15 %. COE then decreases by 11 % from $0.0774/kW-hr to $0.0692/kW-hr.

3.10. Increased Capacity Factor to 90 %

See Appendix A for NETL 's update to capital cost and COE.

In this case, the single-train AHT-2 process configuration remains the same (with process performance remaining the same as in Table 3-17), but the capacity factor increases from 85 % to 90 %. This increased on-stream factor reflects anticipated improvements in process reliability, availability, and maintainability (RAM) resulting from additional operating experience and improvements in control and materials gained through DOE/NETL's demonstration and advanced research programs. As in Section 3.4, it is assumed that these advancements add little additional capital or fixed O&M cost. The increased power production translates into additional revenue, which has a direct positive impact on the COE. Capital and O&M costs for a single- train process are compared in Table 3-20.

The differences between cases lie in variable O&M costs and fuel cost, which increase by about 6 % as the result of increased annual hours of operation. However, the discounted cash flow spreads fixed costs over a greater amount of power production, more than compensating for these additional costs and resulting in an overall decrease in cost of electricity from $0.0761/kWhr to $0.0736/kW-hr – a savings of about 3 % in cost of electricity resulting from increased capacity factor.

Table 3-20. 90 % Capacity Factor: Capital and O&M Cost Summary

	AHT-2 Turbine (Single Train)		90% CF (Single Train)		Δ	
Capital Cost ($1 ,000)						
Plant Sections	**TPC**	**TPC $/kW**	**TPC**	**TPC $/kW**	**Δ TPC $/kW**	**% Δ**
1 Coal and Sorbent Handling	29,100	58	29,100	58	0	0
2 Coal and Sorbent Prep & Feed	47,465	95	47,465	95	0	0
3 Feedwater & Balance of Plant	31,740	63	31,740	63	0	0
4a Gasifier	173,608	346	173,608	346	0	0
4b Air Separation Unit	77,413	154	77,413	154	0	0
5a Gas Cleanup	109,378	218	109,378	218	0	0
5b CO_2 Removal & Compression	19,957	40	19,957	40	0	0
6 Gas Turbine	83,208	166	83,208	166	0	0
7 HRSG	41,401	82	41,401	82	0	0
8 Steam Cycle and Turbines	55,760	111	55,760	111	0	0
9 Cooling Water System	24,243	48	24,243	48	0	0
10 Waste Solids Handling System	34,607	69	34,607	69	0	0
11 Accessory Electric Plant	62,181	124	62,181	124	0	0
12 Instrumentation & Control	21,672	43	21,672	43	0	0
13 Site Preparation	17,652	35	17,652	35	0	0
14 Buildings and Structures	16,184	32	16,184	32	0	0
Total	845,569	1,683	845,569	1,683	0	0
O& M Cost ($1 ,000/yr)						
Fixed Costs	**Total**		**Total**		**Δ**	**% Δ**
Labor	16,535		16,535		0	0
Variable Operating Costs*	**Total**		**Total**			
Maintenance Materials	20,751		21,971		1,220	6
Water	1,110		1,176		66	6
Chemicals	4,565		4,833		268	6
Membrane Replacement	692		732		40	6
Waste Disposal	2,125		2,250		125	6
Total Variable Costs	29,242		30,962		1,720	6
Total O&M Cost	45,777		47,498		1,721	4
Fuel Cost*	57,535		60,920		3,385	6
Discounted Cash Flow Results, levelized						
Capital Cost ($/kW-hr)	0.0396		0.0374			-6
Fixed O&M Cost ($/kW-hr)	0.0051		0.0048			-6
Variable O&M Cost ($/kW-hr)	0.0090		0.0090			0
Fuel Cost ($/kW-hr)	0.0185		0.0185			0
TS&M Cost ($/kW-hr)	0.0039		0.0039			0
Levelized COE ($/kW-hr)	0.0761		0.0736			-3

*Includes 90% Capacity Factor

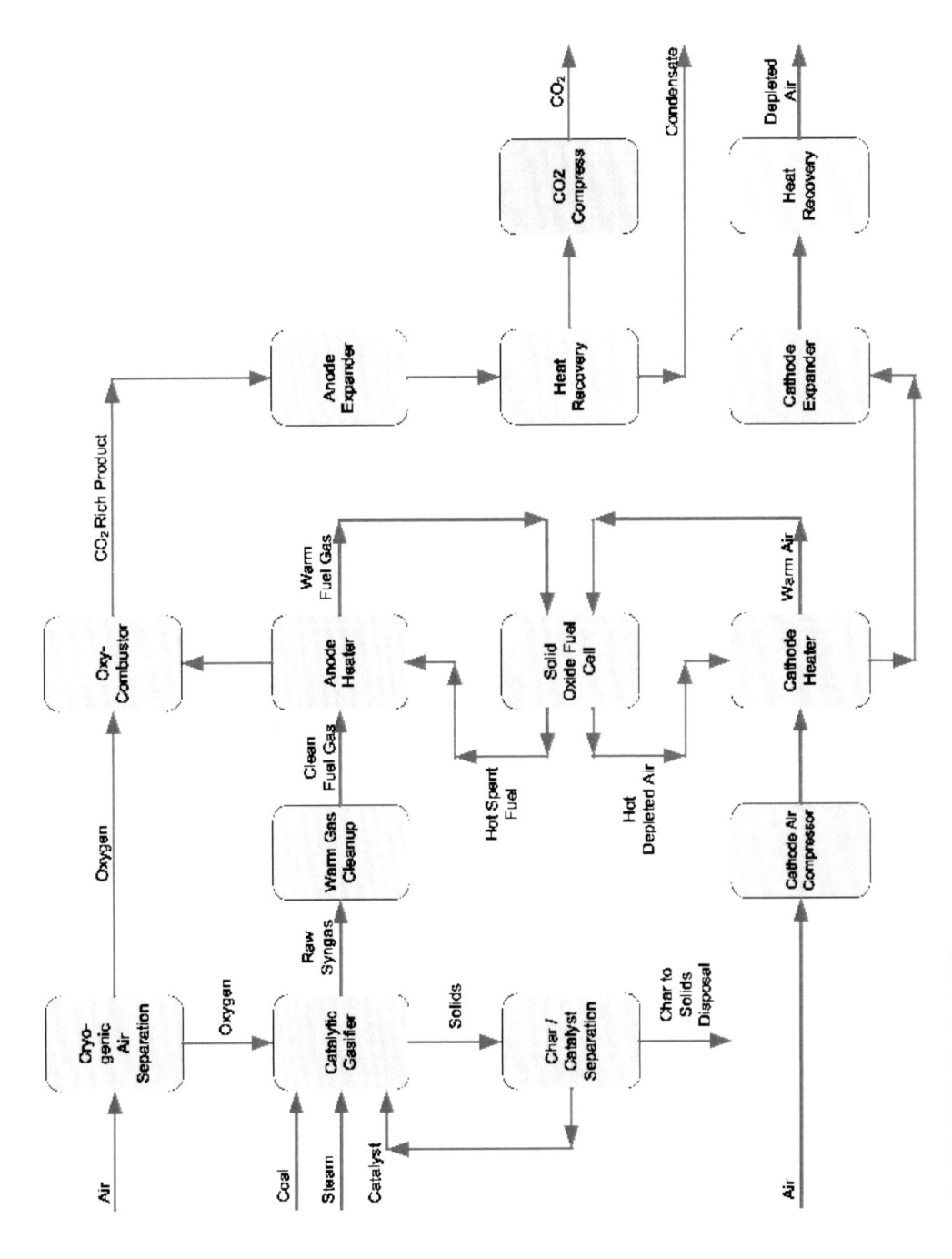

Figure 3-8. Pressurized Solid Oxide Fuel Cell.

Two-Train Configuration

If the capacity factor of the plant having two power trains of AHT-2 turbine is increased from 85 % to 90 %, the COE decreases from \$0.0692/kW-hr to \$0.0671/kW-hr – also a decrease of 3 %.

3.11. Pressurized Solid Oxide Fuel Cell

The IGFC case represents an advanced process configuration that incorporates some, but not all of the advanced technologies in the IGCC pathway. In addition, some advanced conceptual technologies, such as the catalytic gasifier and pressurized oxycombustion unit are added because of their specific value in an IGFC plant.

The non-capture pressurized SOFC process from Volume 1 of this study was modified for carbon capture. This process[6] is ideal for carbon capture because the CO_2-rich fuel cell anode (spent fuel) stream is nearly sequestration-ready. The primary process change is to compress the CO_2 stream to 2,200 psig for transport to storage. A block flow diagram is provided in Figure 3-8. The nominal 600 MW plant size is maintained by adjusting coal feed rate.

Note that even though the CO_2 stream is to be compressed to 2,200 psig, the spent anode stream is still expanded for power recovery. The spent anode stream has 45 % moisture by weight, which is worthwhile to expand in order to recover work from the moisture and then re-compress the CO_2 after removing the moisture.

Another minor process change for this case is to add a bottoming cycle to evaluate the potential for waste heat recovery. The same three-pressure level steam cycle as used in the IGCC cases is used; however due to the larger amount of low quality heat in this case, the exhaust pressure from the low pressure turbine is increased to 1 psia in order to keep the steam quality at about 7 %.

Case Configuration: Catalytic Gasifier, Cryogenic ASU, Warm Gas Cleanup, Solid Oxide Fuel Cell, 90 % Capacity Factor Air

Table 3-21 compares process performance against the non-capture case. Coal feed rate in the carbon capture case increases by 15,000 lb/hr in order to maintain the 600 MW net power output. Increased coal feed rate increases power production from the fuel cell, syngas expander, cathode air expander, and anode exhaust expander. Gross (total) power production increases by 45 MW in the carbon capture scenario.

Auxiliary power use increases in the carbon capture scenario due to (1) additional flow through the cathode air compressor, and (2) need for the CO_2 compressor to pressurize the carbon stream to pipeline pressure.

Net plant efficiency decreases from 59.5 % to 56.3 %. This is a decrease by only 3.2 percentage points, which is less than the 5 percentage point decrease typical of the IGCC cases. Elimination of the need for CO_2 separation in the fuel cell case contributes to improved process efficiency in the carbon capture scenario. Notably, 100 % carbon capture is achieved; there are no carbon emissions other than the CO_2 product stream.

Table 3-21. Comparison of Non-Capture vs. Carbon Capture SOFC Scenario

	Non-Capture SOFC	SOFC With Carbon Capture
Fuel Cell Power (MW)	517	544
Syngas Expander (MW)	22	24
Cathode Air Expander (MW)	208	218
Anode Exhaust Expander (MW)	118	124
Steam Bottoming Cycle (MW)	21	22
Total Power Produced (MW)	886	931
Auxiliary Power Use (MW)	-276	-325
Net Power (MW)	610	606
As-Received Coal Feed (lb/hr)	300,000	315,000
Net Heat Rate (Btu/kW-hr)	5,737	6,063
Net Plant Efficiency	59.5 %	56.3 %
Gasifier Cold Gas Efficiency	92.0 %	92.1 %

Cost Analysis

See Appendix A for NETL's update to capital cost and COE.

Table 3-22 compares the total plant cost, O&M cost, and fuel cost of the non-capture and carbon capture scenarios. A TPC of $700/kW of fuel cell power is assumed for the fuel cell system.[7] The fuel cell system includes fuel cell stack, anode and cathode heaters, anode steam generator and reheat, syngas expander, cathode air compressor, anode and cathode expanders, inverter, catalytic oxidizer and oxygen boost compressor, condensate knockout, and foundations.

Cost accounts in the carbon capture case increase slightly due to the increased coal feed rate and therefore larger equipment sizes. On a $/kW basis, costs of most accounts are similar. The carbon capture case includes a $77/kW cost for CO_2 compression that is not incurred in the non- capture case. Although a larger fuel cell is needed in the carbon capture case, the TPC of the fuel cell decreases by $ 8/kW as the result of a new assumed cost of the fuel cell power island. The CO_2 compressor accounts for most of the $ 127/kW net increase in cost for the carbon capture process. This fuel cell case represents much less of an increase in TPC for the carbon capture scenario than any of the IGCC cases; the sequestration-ready CO_2 stream exiting the fuel cell accounts for the avoidance of increased cost for CO_2 separation in the gas cleanup account.

4. SUMMARY OF ADVANCED TECHNOLOGY IMPROVEMENTS

The information presented in the previous section is consolidated in the following discussion in order to summarize the relative benefits of the advanced technologies in both non-capture and carbon capture scenarios.

Table 3-22. SOFC: Capital and O&M Cost Summary

	Non-Capture SOFC		SOFC With Carbon Capture		Δ	
Capital Cost ($1,000)						
Plant Sections	TPC	TPC $/kW	TPC	TPC $/kW	Δ TPC $/kW	% Δ
1 Coal and Catalyst Handling	30,814	51	31,764	52	1	2
2 Coal and Catalyst Prep & Feed	41,428	68	42,817	71	3	4
3 Feedwater & Balance of Plant	21,649	35	22,119	36	1	3
4a Gasifier	155,335	255	160,426	265	10	4
4b Air Separation Unit	81,306	133	84,494	139	6	5
5a Gas Cleanup	66,351	109	77,449	128	19	17
5b CO_2 Removal & Compression	0	0	46,376	77	77	∞
6 Gas Turbine	0	0	0	0	0	0
7 Fuel Cell	387,875	636	380,780	628	-8	-1
8 Steam Cycle and Turbines	15,542	25	16,073	27	2	8
9 Cooling Water System	13,711	22	14,079	23	1	5
10 Waste Solids Handling System	35,692	59	36,782	61	2	3
11 Accessory Electric Plant	88,137	144	92,989	153	9	6
12 Instrumentation & Control	28,911	47	30,282	50	3	6
13 Site Preparation	18,823	31	19,149	32	1	3
14 Buildings and Structures	10,097	17	10,241	17	0	0
Total	995,670	1,632	1,065,820	1,759	127	8
O&M Cost ($1,000/yr)						
Fixed Costs	Total		Total		Δ	% Δ
Labor	19,542		21,045		1,503	8
Variable Operating Costs*	Total		Total			
Maintenance Materials	28,487		29,552		1,065	4
Water	168		354		186	111
Chemicals	3,845		3,950		105	3
Fuel Cell Stack Replacement	17,835		18,759		924	5
Waste Disposal	2,397		2,481		84	4
Total Variable Costs	52,731		55,096		2,365	4
Total O&M Cost	72,273		76,141		3,868	5
Fuel Cost*	49,799		52,289		2,490	5
Discounted Cash Flow Results, levelized						
Capital Cost ($/kW-hr)	0.0362		0.0390			8
Fixed O&M Cost ($/kW-hr)	0.0047		0.0051			9
Variable O&M Cost ($/kW-hr)	0.0127		0.0133			5
Fuel Cost ($/kW-hr)	0.0124		0.0132			6
TS&M Cost ($/kW-hr)	NA		0.0039			∞
Levelized COE ($/kW-hr)	0.0661		0.0745			13

*Includes 90 % Capacity Factor

4.1. Process Efficiency

The following Figure 4-1 shows the cumulative improvement in process performance as each technology is introduced to the composite process. The uppermost curve represents non-capture scenarios, which consistently have higher process efficiency than the carbon capture scenarios. Cases that feature improved capacity factor do not affect performance efficiency because the capacity factor merely increases the percentage of on-stream operation.

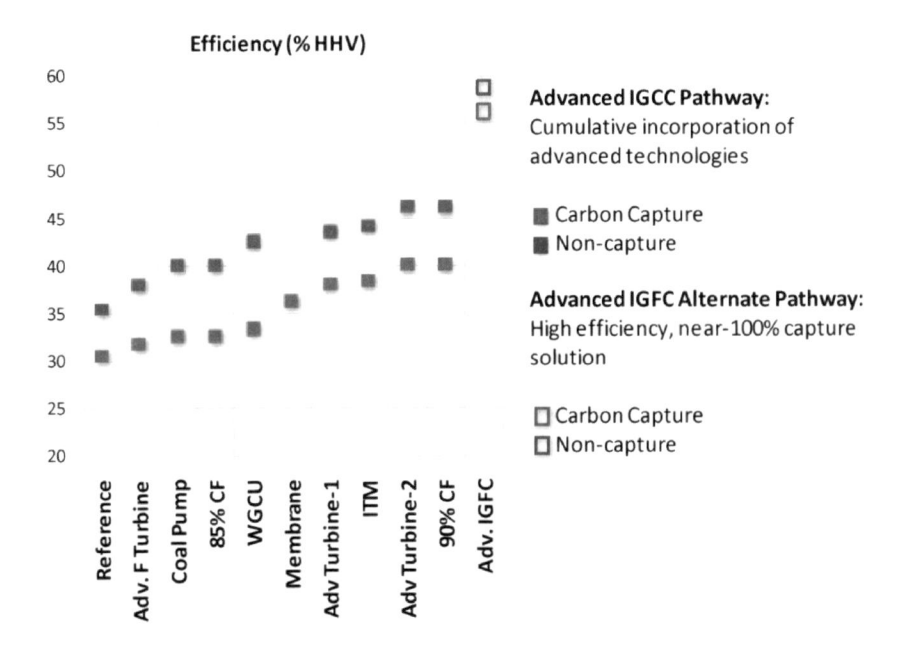

Figure 4-1. Cumulative Impact of R&D on Process Efficiency.

Advanced turbines contribute strongly to increased process efficiency due to the combination of improved engine performance at increasingly higher pressure ratios and firing temperatures, and also increased turbine exit temperature, which improves heat recovery from the HRSG – especially if an increase in steam superheat temperature is involved. The 1.3 percentage point (%pt) improvement of the advanced "F" turbine is not as great in a carbon capture scenario as it is in the non-capture scenario (2.5 %pt); air integration is not possible in the carbon capture scenario, and the turbine exit temperature is not high enough that steam superheat temperature can be increased. When the first generation advanced turbines are introduced, however, the efficiency of the carbon capture scenario increases (1.8 %pt) more than in the non-capture scenario (1.0 %pt); this is due to the additional contributions of air integration and increased steam superheat temperature. The next-generation advanced turbines (Adv Turbine-2) contribute 2.0 and 1.7 %pt improvements to the non-capture and carbon capture scenarios, respectively. The total performance improvement due to the advanced turbines, therefore, is 5.5 %pt in the non-capture scenario and 4.8 %pt in the carbon capture scenario.

The coal feed pump makes a greater contribution to process efficiency improvement in the non- capture scenario (2.1 %pt) than in the carbon capture scenario (0.8 %pt). The coal feed pump increases process efficiency by eliminating the need to evaporate water in a slurry-fed gasifier. In the non-capture scenario with cold gas cleanup, that moisture is condensed and most of the latent heat is unrecoverable because of the low condensation temperature. In the carbon capture scenario with cold gas cleanup, on the other hand, moisture is needed for sour shift; so whether the moisture is provided by slurry water or addition of shift steam (following a dry feed gasifier) doesn't have as much of an impact on process efficiency.

Warm gas cleanup (with Selexol CO_2 capture) improves process efficiency over cold gas cleanup by 0.8 %pt in the carbon capture scenario as the result of eliminating the sour water stripper reboiler duty; the improvement is not as great as the 2.5 %pt increase in the non-capture scenario because syngas is quenched prior to Selexol, knocking moisture out of flue gas that otherwise remains in the turbine fuel in the non-capture case – providing added flow through the turbine. However, warm gas cleanup (with hydrogen membrane) contributes an additional 2.9 %pt in process efficiency in the carbon capture scenario by eliminating the Selexol reboiler and auxiliary power, and also producing CO_2 at elevated pressure – reducing CO_2 compressor load.

The ITM does not contribute strongly to process performance in either the non-capture or carbon capture scenarios. The primary benefit of the ITM, as will be seen in the following discussion, is decreased capital cost of oxygen production.

Overall, advanced technologies increase IGCC process efficiency by as much as 10.7 %pt in non-capture scenarios and by 9.3 %pt in carbon capture scenarios. Non-capture scenarios benefit from (1) greater percentage of air integration for each turbine model due to the difference in syngas versus hydrogen fuel flow; (2) reduced coal flow rate per unit net power generation, thus reducing parasitic load of oxygen production; (3) no need for shift steam generation, thus increasing steam turbine power generation, and (4) no need for CO_2 compression, thus reducing parasitic losses.

The pressurized solid oxide fuel cell cases – both capture and non-capture – are capable of process efficiencies that approach 60 %. The catalytic gasifier, with high methane content in the syngas, operates with a cold gas efficiency in excess of 90 %. Conversion of chemical energy within the fuel cell, as opposed to thermal and mechanical energy in an IGCC process, enables the higher process efficiencies obtained in the SOFC cases. The difference in process efficiency between the non-capture and capture scenarios is simply due to the power needed to compress CO_2 to pipeline delivery pressure.

4.2. Total Plant Cost

See Appendix A for NETL's update to capital costs. [8]

As each advanced technology is introduced to the composite process, total plant cost generally decreases as shown in Figure 4-2. The uppermost curve represents the carbon capture scenarios, which consistently have higher TPC due, at a minimum, to (1) additional equipment needed for CO_2 separation and compression; (2) additional equipment needed for shift steam generation, and (3) reduced net power generation. Improved capacity factor has no effect on TPC, as seen in Figure 4-2, just as it has no effect on process efficiency.

Advanced gas turbines significantly reduce total plant cost. Although the cost of the turbine itself increases due to increased size, TPC on a $/kW basis decreases because of increased net plant power. As in the discussion above on process efficiency, the advanced "F" turbine has more impact ($304/kW) in the non-capture scenario (versus $246/kW) because of air integration and increased steam superheat temperature. The carbon capture case catches up somewhat when air integration and increased superheat temperature are introduced with the AHT-1 turbine; the non-capture cost reduction is $72/kW compared to $192/kW with carbon capture. As discussed in Section 3, the impact of the next-generation of advanced turbines is diminished by economy of scale when the number of trains is reduced from two to one in order to maintain the nominal 600 MW plant size; the TPC reductions are $27/kW and $41/kW for the non-capture and carbon capture scenarios, respectively. The bottom of the shaded bars in Figure 4-2 indicate that TPC continues to decrease if two trains turbine trains are installed – doubling the plant output and decreasing TPC by $219/kW in the non-capture scenario and by $254/kW in the carbon capture scenario.

The coal feed pump has negligible impact on TPC in a carbon capture scenario – only $7/kW compared to the $60/kW reduction in the non-capture scenario. This is because of the minor cost of equipment, coupled with greater reduction in net plant power (due to need for shift steam generation) in the carbon capture scenario than in the non-capture scenario.

While warm gas cleanup results in greater process efficiency improvement for the carbon capture scenario as shown above in Figure 4-1, its impact is especially pronounced in terms of TPC. The cost of warm gas desulfurization is less than single-stage Selexol to begin with (which partly accounts for the decrease in TPC of the WGCU+Selexol non-capture and carbon capture scenarios in Figure 4-2), but when the cost savings from eliminating the second stage Selexol absorber for CO_2 capture is added, the decrease in TPC of the gas cleanup section for the WGCU+Membrane carbon capture scenario becomes much greater. The cost of CO_2 compression, likewise, is much less in the WGCU+Membrane case than any of the previous carbon capture cases due to the higher pressure at which CO_2 is produced from the H_2 membrane. Finally, when the added net power generation (made possible by eliminating sour water stripper and Selexol reboiler duties and reduced CO_2 compression parasitic loss) is divided into the already-reduced TPC, the cost of the warm gas cleanup cases on a $/kW basis become $40/kW (for WGCU+Selexol) and $418/kW (for WGCU+Membrane) less than the cold gas cleanup carbon capture scenario.

The ITM is seen to reduce TPC by relatively more in the carbon capture scenario ($131/kW) than in the non-capture scenario ($82/kW). With increase in coal feed rate to generate hydrogen turbine fuel as opposed to syngas turbine fuel, the significance of the air separation unit increases. In other words, with increased oxygen demand in the carbon capture cases, the capital cost savings represented by the less-expensive ITM compared to cryogenic ASU has a greater impact on reducing cost.

Overall, a capital cost reduction of about $700/kW is anticipated from advanced technologies in non-capture IGCC applications. Even more significant, however, is an anticipated $1,000/kW reduction in cost for carbon capture IGCC applications.[9] The primary reasons for greater TPC reductions in the carbon capture scenarios are: (1) low cost of H_2 membrane for advanced CO_2 separation technology; (2) reduced parasitic load of CO_2 compression (and therefore increased net plant power generated) due to high pressure at which CO_2 is separated by the H_2 membrane, and (3) reduced cost of CO_2 compressor equipment, again because of high pressure CO_2 separation.

The TPC of the most advanced IGCC process with carbon capture is nearly $280/kW greater than its non-capture counterpart. The SOFC capital cost, on the other hand, increases by only about $ 130/kW when carbon capture is added; the incremental cost to the SOFC scenario is essentially the CO_2 compressor, which is a relatively minor impact compared to the IGCC scenarios. The TPC of the carbon capture SOFC scenario is slightly greater than the most advanced IGCC configuration with carbon capture ($ 1,759/kW versus $ 1,683/kW).

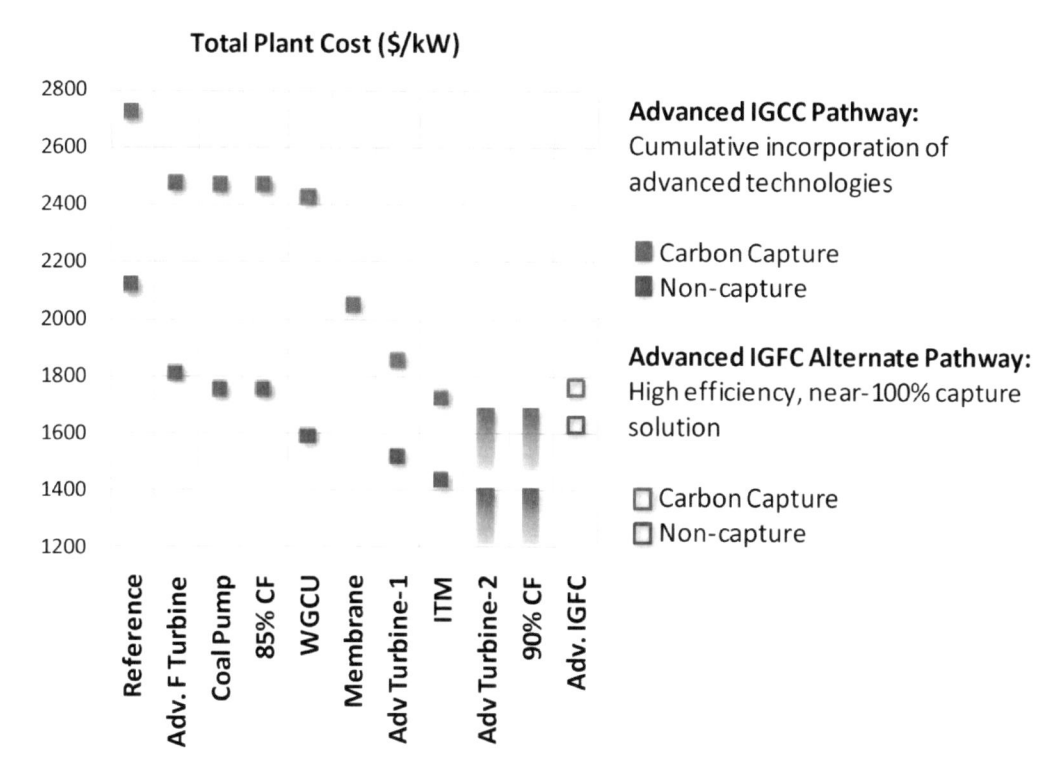

Figure 4-2. Cumulative Impact of R&D on Total Plant Cost.

4.3. Cost of Electricity

See Appendix A for NETL 's update to COE.

As each new advanced technology is step-wise implemented in the advanced power system, the reduction in COE is represented in Figure 4-3. Effects of improved capacity factor become as significant as the other technology improvements that yield increased process efficiency and decreased capital cost. The increase to 85 % capacity factor results in a 4 % reduction in COE for both the non-capture and the carbon capture scenarios. The increase to 90 % capacity factor results in an additional 3 % reduction in COE for both the non-capture and carbon capture scenarios.

The advanced "F" turbine and the AHT-1 turbine contribute significant COE reductions in carbon capture scenarios – by 8.4 mills/kW-hr and 6.6 mills/kW-hr, respectively. The reduction in COE is slightly greater than in the non-capture scenarios (13.4 mills/kW-hr total). Due to economy of scale, the nominal 600 MW plant with a single AHT-2 turbine train results in a small (1.3 mills/kW-hr) decrease in COE. If two process trains are used as in the other IGCC plants, however, COE decreases by 12 % in the non-capture scenario and by 11 % in the carbon capture scenario.

Consistent with no appreciable change in either process efficiency or TPC, the coal feed pump has little impact on COE in an IGCC process with carbon capture.

Warm gas cleanup has a much greater impact on carbon capture IGCC scenarios than on the non- capture scenarios; this is chiefly due to the large decrease in TPC resulting from CO_2 separation and compression and increased net power generation. In the case in which warm gas cleanup is introduced together with the H_2 membrane, COE decreases by 13.4 mills/kW-hr or 13 % compared to cold gas cleanup.

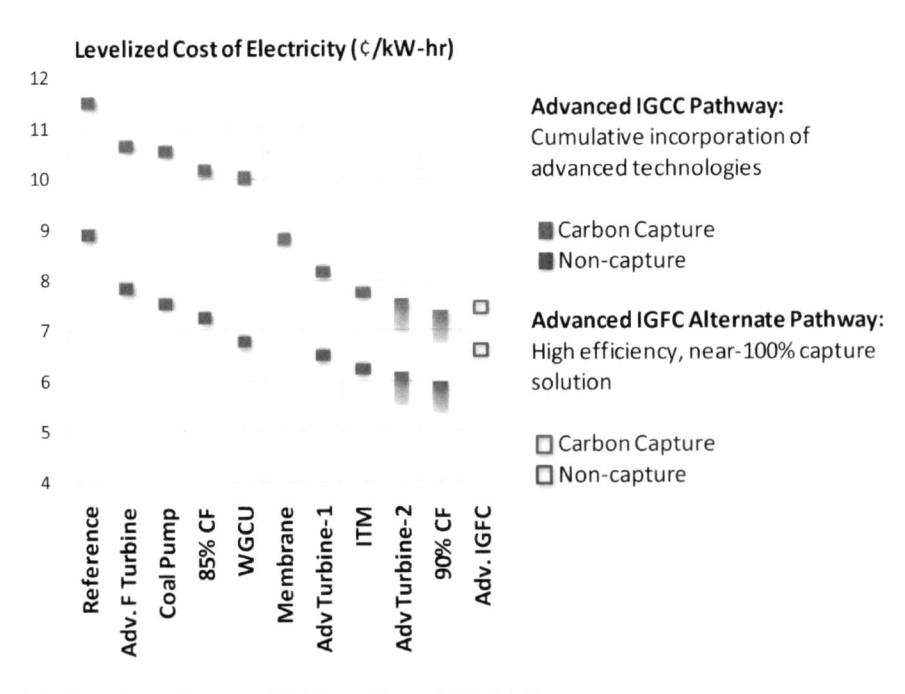

Figure 4-3. Cumulative Impact of R&D on Cost of Electricity.

ITM technology decreases the COE by 4.0 mills/kW-hr in the carbon capture scenario. It has a more pronounced effect on carbon capture scenarios than non-capture because, as explained above, coal feed rate increases for the carbon capture cases, providing more opportunity for cost reduction in the ASU. By comparison, the COE reduction in the non-capture scenario is 2.6 mills/kW-hr.

For a nominal 600 MW plant, cumulative reductions in COE resulting from advanced technology are 29 mills/kW-hr for non-capture IGCC scenarios, but 41 mills/kW-hr for carbon capture IGCC scenarios. Advanced technology, therefore, represents 23 % and 36 % reductions in COE for non-capture and carbon capture scenarios, respectively.

COE in the non-capture SOFC scenario increases by 11 % over that of the most advanced non- capture IGCC technology; this is due to a higher TPC that, even despite much higher process efficiency, results in a COE that is greater than IGCC by 6.6 mills/kW-hr. In the carbon capture scenario, the sequestration-ready CO_2 stream incurs minimal incremental capital cost for carbon capture. The resulting COE, aided by very high process efficiency, is 0.9 mills/kW-hr greater than the most advanced IGCC configuration with carbon capture.

4.4. DOE's Carbon Capture Targets

DOE's advanced power generation program goals are to achieve 90 % carbon capture while maintaining less than 10 % increase in COE over a 2003 reference IGCC plant having no carbon capture. That reference plant is represented in Case 0 in Volume 1 of this study. It consists of a slurry-fed gasifier, cryogenic ASU, single stage Selexol for sulfur removal, and 7FA syngas turbine. At 75 % capacity factor the COE of that plant is 9.3 ¢/kW-hr, so DOE's cost target for carbon capture is 10 % greater, or 10.2 ¢/kW-hr.

From Figure 4-3, DOE's carbon capture target will be met early in the pathway, specifically by the case with 85 % capacity factor. Other features of that case include advanced "F" hydrogen turbine, dry feed gasifier, cryogenic ASU, and cold gas cleanup.

All subsequent technology advancements will help to exceed DOE's carbon capture targets. By achieving the ultimate, most advanced IGCC and IGFC technologies projected in Figure 4-3, DOE could realize a 20 % *reduction* in COE over a 2003 IGCC plant having no carbon capture. The enabling technologies to achieve that improvement include:

- Advanced hydrogen turbines
- Warm gas cleanup
- Pressurized SOFC with catalytic gasifier Improved RAM
- ITM
- Coal feed pump

The technology pathway evaluated in this study covers a time span of about eighteen (18) years of technology development. Results of the analysis clearly indicate the importance of continued R&D, large scale testing, and integrated deployment so that future coal-based power plants will be capable of generating clean power with greater reliability and at significantly lower cost.

Aside from improved process efficiencies and reduced costs of electricity for both non-capture and carbon capture power generation alike, these advanced technologies enable (1) production of high-value products such as hydrogen, (2) integration with solid oxide fuel cells, and (3) precombustion carbon capture projected at lower cost than post-combustion alternatives.

APPENDIX A: NETL UPDATE TO COST REPORTING

Revision 1 of the NETL Baseline Study [2] served as the primary basis for the performance and cost of conventional technology components in this report and provided the financial structure and cost of electricity calculation methodology. Revision 2 of the NETL Baseline Study (Nov 2010) [5] updates performance and significantly revises the reporting of capital costs and costing methodology. This Appendix provides the estimated capital cost and COE for each of the cases presented in this report consistent with the cost modifications in Revision 2 of the Baseline Study.

Summary of Modifications

Revision 2 of the NETL Baseline Study included (1) performance/simulation updates, and (2) multiple changes to costs and cost reporting bases.

Performance Changes

Revision 2 performance modeling changes for Case 2 in the Baseline Study have the potential to improve the performance of the corresponding case in this report (Adv "F" Turbine).[10] However, it is not yet known if those improvements would translate into improvements for all subsequent advanced technology cases in this report. To address this discrepancy, this appendix modifies the efficiencies as follows: (1) the Adv "F" Case efficiency was set equal to that of Case 2 in Revision 2 of the NETL Baseline Study, (2) the efficiency of the most advanced IGCC case of 40.0% was maintained consistent with this study, and (3) the efficiencies of all intermediate cases were proportionally adjusted. This results in a slight reduction in the incremental efficiency improvements for each cumulative addition of advanced technology. No change was made to the efficiency of the advanced IGFC configuration.

Key Cost and Cost Reporting Modifications

Capital costs in Revision 2 of the NETL Baseline Study were reassessed at a component-bycomponent level. Updates to the capital costs in this report were revised and estimated at the plant level. A more detailed component-by-component level revision is planned for future revisions.

The remaining changes to Revision 2 of the Bituminous Baseline report that have been incorporated into the results presented in this appendix are as follows:

- All costs are reported in June 2007 dollars. June 2007 capital costs are approximately equal to January 2010 costs based on the *Chemical Engineering* Plant Cost Index.
- Previously excluded capital costs, such as owner's costs, have been added and are reported as Total Overnight Cost (TOC). Costs are also presented as Total As-Spent Cost (TASC). Figure A-1 provides additional detail on what is included at each cost level.
- The COE now includes owner's costs, and interest and escalation during construction.
- The bituminous coal cost used in this study is $1.64/MMBtu. This cost was derived from data in the Energy Information Administration Annual Energy Outlook.
- Property taxes and insurance have been included as part of the fixed O&M cost.
- CO_2 TS&M costs have been updated.
- All O&M costs, including fuel, are assumed to escalate at a nominal rate of 3%, consistent with the assumed inflation rate.
- The operation period assumed for levelization is 30 years. The capital expenditure period is 5 years (one year of capital expenditure prior to construction and four years of construction).
- LCOE continues to be based on a current-dollar analysis, but the levelization factor calculation has been modified.

Summary of Modified Results

Figure A-2 depicts the cumulative improvements in process efficiency, TOC, and first-year COE as each technology is introduced for the carbon capture cases described in this study and the non- capture cases from Volume 1. TOC and first-year COE are updated consistent with the changes to Revision 2 of the NETL Baseline Study described above.

The bottom of the shaded bars on the TOC and COE pathways illustrate the impact of the AHT-2 turbine if two turbine trains were built. That installation would exceed the nominal 600 MW plant size for this study, but the point serves to illustrate the effect of economy of scale on process economics.

Table A-1 summarizes the updated results for each case with CCS.

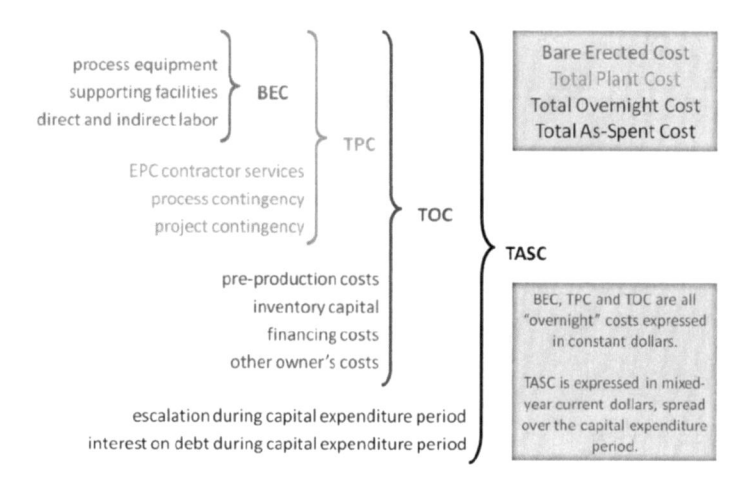

Figure A-1. Elements of Capital Costs.

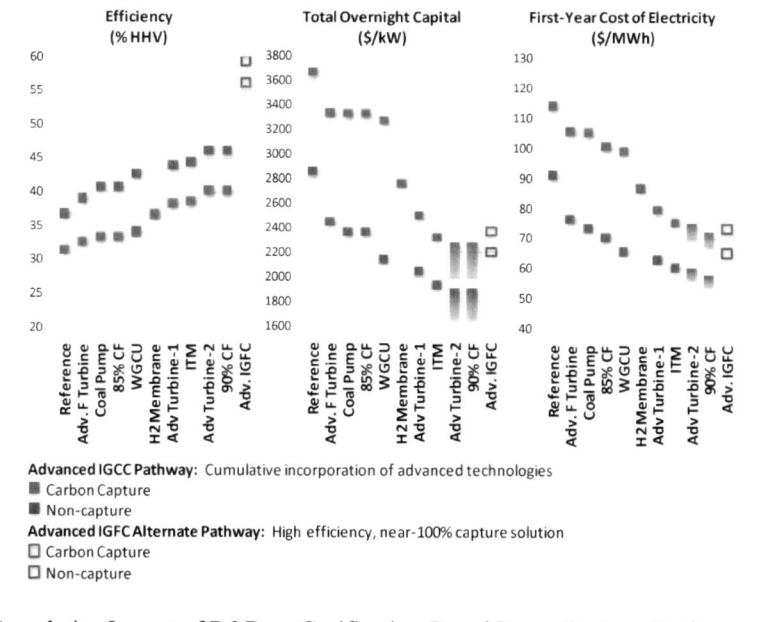

Figure A-2. Cumulative Impact of R&D on Gasification-Based Power Systems Performance and Cost

Table A-1. Summary of Updated Capital Costs and Cost of Electricity

All costs in June 2007 dollars (≈January 2010 dollars) unless otherwise indicated	Reference IGCC with CCS	Adv "F" Turbine	Coal Feed Pump	85% CF	WGCU/Selexol	WGCU/H_2 Membrane	AHT-1 Turbine	ITM	AHT-2 Turbine	90% CF	Advanced IGFC with CCS
HHV Efficiency, %	31.5%	32.6%	33.3%	33.3%	34.0%	36.6%	38.2%	38.5%	40.0%	40.0%	56.3%
Net Plant Output, MW	444	543	510	510	535	572	659	691	502	502	606
Capacity Factor / Availability	80%	80%	80%	85%	85%	85%	85%	85%	85%	90%	90%
TPC, $/kW	2,980	2,710	2,700	2,700	2,660	2,240	2,030	1,890	1,850	1,850	1,930
TOC, $/kW	3,670	3,330	3,330	3,330	3,270	2,760	2,500	2,330	2,270	2,270	2,370
TASC, $/kW (mixed year dollars)	4,180	3,800	3,790	3,790	3,730	3,150	2,850	2,650	2,590	2,590	2,700
30-Year Levelized[1] COE, $/MWh	145	134	133	128	126	110	101	96	95	91	93
COE[2], $/MWh	114	106	105	101	99	87	80	76	75	72	73
Capital	65	59	59	56	55	46	42	39	38	36	37
Fixed O&M	15	15	15	14	13	12	11	10	10	9	10
Variable O&M	10	9	9	9	10	9	9	8	9	9	13
Fuel	18	17	17	17	16	15	15	15	14	14	10
CO_2 TS&M	6	5	5	5	5	5	4	4	4	4	4
Cost of Avoiding CO_2 [2], $/tonne Relative to Supercritical PC without CCS	78	66	65	58	56	39	29	23	22	18	18

[1] Current-dollar levelization

[2] Assumes 3% nominal escalation per year of COE, fuel cost and O&M cost over the 30-year capital recovery period

LIST OF REFERENCES

[18] "Current and Future IGCC Technologies. (October 16, 2008). *A Pathway Study Focused on Non-Carbon Capture Advanced Power Systems R&D Using Bituminous Coal – Volume 1.*" Department of Energy, National Energy Technology Laboratory. DOE/NETL2008/1337.

[19] "Cost and Performance Baseline for Fossil Energy Plants. (1, August, 2007). Volume *1*: Bituminous Coal and Natural Gas to Electricity." Revision. Department of Energy, National Energy Technology Laboratory. DOE/N ETL-2007/1 281_r1.

[20] "IGCC: What's GE Up To?" (October 13, 2005). Norm Shilling, General Electric. American Coal Council 2005 Coal Market Strategies.

[21] "The Benefits of SOFC for Coal-Based Power Generation." (October 30, 2007). Report Prepared by E. Grol, J. DiPietro, and J. Thijssen for Wayne Surdoval. National Energy Technology Laboratory

[22] "Cost and Performance Baseline for Fossil Energy Plants. (2, November, 2010). Volume *1*, Bituminous Coal and Natural Gas to Electricity." Revision. Department of Energy, National Energy Technology Laboratory. DOE/NETL-201 0/1 397.

End Notes

[1] NETL is updating the performance, cost, and costing methodology as part of Revision 2 of "Cost and Performance Baseline for Fossil Energy Plants, Volume 1: Bituminous Coal and Natural Gas to Electricity." The estimated capital cost and COE for the configurations presented in this report using this new methodology are reported in Appendix A.

[2] The pseudonyms AHT-1 and AHT-2 are used to represent technology that is presently under development within DOE's R&D program. While actual performance parameters are business-sensitive, the turbine parameters used in this study represent target performance.

[3] NETL is updating the performance, cost, and costing methodology as part of Revision 2 of "Cost and Performance Baseline for Fossil Energy Plants, Volume 1: Bituminous Coal and Natural Gas to Electricity." The estimated capital cost and COE for the configurations presented in this report using this new methodology are reported in Appendix A.

[4] Warm gas cleanup chemical costs were verified by personal communication with Brian Turk, RTI.

[5] Detailed models of hydrogen turbines were not developed for this study. As such, the power rating of each hydrogen turbine model is assumed to be the same as the corresponding syngas turbine.

[6] The pressurized SOFC process proposed by SAIC in the NETL report titled "The Benefits of SOFC for Coal-Based Power Generation" prepared by E. Grol, J. DiPietro, and J. Thijssen dated October 30, 2007 is the basis of this process design. [5]

[7] The assumed cost of the fuel cell has changed since Volume 1.

[8] NETL is updating the performance, cost, and costing methodology as part of Revision 2 of "Cost and Performance Baseline for Fossil Energy Plants, Volume 1: Bituminous Coal and Natural Gas to Electricity." The estimated capital cost and COE for the configurations presented in this report using this new methodology are reported in Appendix A.

[9] TPC reduction is $1 ,000/kW for a nominal 600 MW-size plant (single A HT-2 turbine train); the reduction in TPC becomes $1 ,235/kW if two trains of AHT-2 turbine are built.

[10] Revision 1 of the NETL Cost and Performance Baseline Volume 1 assumed "free" recovery of hydrogen and other components from the CO_2–rich streams exiting Selexol. Revision 2 modified the Selexol performance to correspond to a high hydrogen recovery, eliminating any need for further purification of the CO_2 streams exiting Selexol. This performance change was already incorporated in the initial publication of the corresponding cases in this report.

INDEX